T0322187

Springwatch

BBC

Birdtopia

Birdtopia

A Miscellany of British Birds

Jack Baddams

1

BBC Books, an imprint of Ebury Publishing
20 Vauxhall Bridge Road
London SW1V 2SA

BBC Books is part of the Penguin Random House group of companies whose addresses can be found at global.penguinrandomhouse.com

This book is published to accompany the television series *Springwatch* broadcast on BBC Two. *Springwatch* is a BBC Studios production.

Executive Producer: Rosemary Edwards
Series Editor: Joanna Brame
Series Producer: Laura Whitley

First published by BBC Books in 2024

www.penguin.co.uk

A CIP catalogue record for this book is available from the British Library

ISBN 9781785948688

Commissioning Editor: Phoebe Lindsley
Design: seagulls.net
Production: Antony Heller
Illustrations: David Wardle

Printed and bound in Great Britain by Clays Ltd, Elcograf S.p.A.

The authorised representative in the EEA is Penguin Random House Ireland, Morrison Chambers, 32 Nassau Street, Dublin D02 YH68.

Contents

Introduction

Does any country love its birds quite like Britain does?

OK, perhaps a difficult question to answer, but our affinity for the feathered is strong and stretches back deep into the past, evident in the lives of all those who have ever called our islands home.

From the bones of Eagles found buried alongside our ancestors in sacred tombs, to the presence of the mythical 'Martlets' on the royal arms of Anglo-Saxon kings, it is clear that birds have played important roles in our lives for millennia.

Delve into British literature and birds leap from the page in abundance, having caught the imaginations of our most famed poets, songwriters and story tellers. Shakespeare, Chaucer, Keats, Wordsworth, Burns – the list of wordsmiths who have included birds in their works is testament to the feelings they inspire in us.

Today, we have a national obsession with feeding our garden birds, one unmatched by any other country on the planet. It's estimated that over half of UK households put out food for the birds, providing more than 150,000 tonnes of bird food per year. To put that in perspective, that's enough food to maintain the entire combined populations of our most common bird feeder-using species THREE TIMES OVER!

Guillemot

Eider duck

In short, we love our wild birds. It's therefore no surprise that a programme like *Springwatch* has become such a staple in the TV schedule. Since its beginnings in 2005, the show has given viewers a unique, intimate insight into the lives of the species that live alongside us.

By using small, remotely operated cameras, and a dedicated team watching them 24 hours a day, *Springwatch* has lifted the lid on the dramas that are taking place in hedgerows, woodlands, reedbeds and fields across the country each spring.

From Blue Tits to Bitterns, we've watched as chicks have developed and then celebrated when the characters we've grown to love have taken their first flights into the big wide world. Not all will make it, however, and, whether it be the effects of bad weather or a hungry predator, *Springwatch* has never shied away from showing the challenges that birds face when it comes to raising a family.

This book is a journey through the species that share our home. We'll revel in the facts, folklore and fiction that surround them, our ornithological exploration taking us from ancient fables to brand new science as we seek to get under the feathers of Britain's birds. In doing so, we will revisit some of *Springwatch*'s most famous characters from over the years and look back on some of the unique behaviours that have played out in front of our cameras.

And the best thing about it all?

We're not talking about birds in some far off land, on the other side of the world, in tropical rainforests or exotic savannahs. These are the ones we can see whenever we step outside our doors. They're right there, for everyone.

Get out and enjoy them.

Razorbill

Swans

Chances are, if you conducted a survey of the British public and asked them what they knew about swans, you'd get two answers:

1) They're all owned by the King.
2) They can break your arm with their wings.

The second statement is a complete myth. Although swans are heavy, powerful birds that are fiercely protective of their young, they lack the strength to be able to snap an ulna. The first statement, however, is *mostly* true.

There are three species of swan that can be found naturally in the UK. The Mute Swan is the most common and will be familiar to almost everyone, with its bright orange beak making it very distinctive. Whooper and Bewick's Swans only visit our isles in the winter, to escape the freezing temperatures of the northern climes where they breed. These two species have black and yellow bills, with the Bewick's being a smaller, more compact version of its larger Whooper cousin.

Mute swan

SOVEREIGN SWANS

It's just the one swan species that is subject to the royal decree. The ruling monarch of the UK has the right to claim any unmarked Mute Swan on public lands and the first record of the Crown's claim can be dated back to 1186. In the past, swans were an important source of food. These large birds provided lots of meat and weren't particularly hard to hunt. They were favoured by the wealthy, often being served up at banquets. In 1247, Henry III was believed to have ordered 40 swans for his Christmas feasts.

Long gone are the days when swans were served up on the dinner table, and King Charles III, the current 'Seigneur of the Swans', only practises ownership over parts of the River Thames and its tributaries. Every July, one of Britain's most obscure traditions takes place when the Royal Family's official Swan Marker leads a team of 'Swan Uppers' out onto a 79-mile stretch of the Thames to conduct a census of the King's flock.

Adult birds and their cygnets are rounded up to have their health checked and make sure they are free of any injury caused by fishing line or other such hazards. Cygnets are ringed if their parentage denotes they belong to either the Vintners or the Dyers livery companies. These companies were granted rights by the Crown to own swans for themselves in the fifteenth century. Exactly why they were given this privilege isn't known, but it's believed likely to be a way of strengthening relationships between the Royals and powerful trade guilds. All birds belonging to the Crown are left unmarked and all are returned to the river shortly after their capture.

HIGHLY PROTECTED

The importance and status of swans were reflected in the punishments for interfering with them. A statute dating from Henry VII's reign proclaimed that 'anyone stealing or taking a swan's egg would have one year's imprisonment and make payment of a fine at the King's will'. The penalty for killing one in Germany was a fine of as much corn as would be able to cover a swan's corpse whilst its beak was being pulled up to its maximum extent.

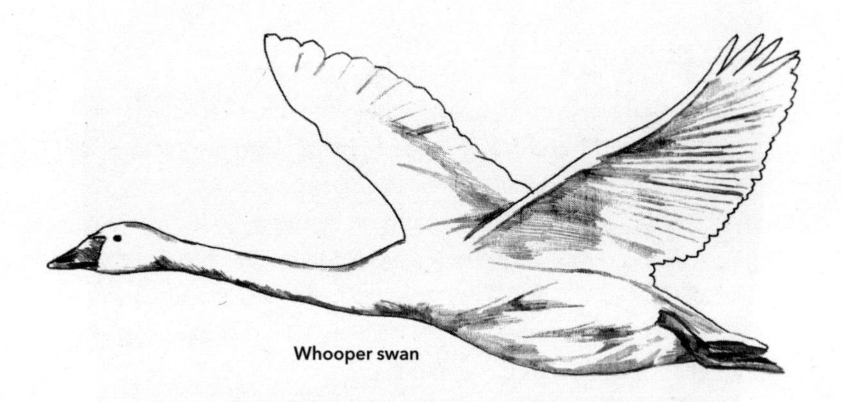

Whooper swan

A FAVOURITE THROUGH THE AGES

Representations of swans are found throughout human history. The earliest are carvings made from the ivory of mammoths that were found in Siberia. They depict swans flying, with their necks outstretched in front of them, and date back over 20,000 years to the last Ice Age. The arrival of large migratory birds like swans would have been important seasonal markers for hunter-gatherer communities in the past and they may have been celebrated with special ceremonies. Our evidence of humans interacting with swans may date as far back as 420,000 years thanks to a swan bone found in Qesem Cave in Israel that bears cut marks that seem to have been done in a specific way so as to remove the long flight feathers.

SINGING SWANS

By the time of Ancient Greece, the phrase 'swan song' was in use to describe a final gesture, effort or performance before death or retirement. Its origin comes from the belief that swans, after being silent all their lives, would sing a beautiful song just before their death. It's a myth that's been challenged ever since it emerged, with the Roman author Pliny the Elder writing, 'Observation shows that the story that the dying swan sings is false.'

Despite its name, the Mute Swan isn't actually silent and uses a series of grunts and, as anyone who has ever walked a dog too close to a swan will know, aggressive hisses to communicate. Both Whooper

Bewick's swan

and Bewick's Swans are much more vocal, possessing lovely bugling calls that they use to enthusiastically communicate with each other, particularly in flight when they're migrating thousands of miles.

TILL DEATH DO US PART

Swans are famed for their strong pair bonds and are certainly one of our most faithful types of bird. Most other species will form short term pairs that last a single breeding season and, even then, there is a high chance that some chicks in the nest will not belong to the male bird tending to them. Swans take a different approach and, once a pair bond is cemented, they will rarely separate. Whooper and Bewick's Swans will migrate thousands of miles together, whilst studies looking at Mute Swans have shown that successfully breeding pairs have a divorce rate of around only 3%. With a potential lifespan of 30 years, Swan pairs can remain together for a very long time and are only usually separated by death. It is however a myth that birds who have lost their mates choose to remain single for the rest of their lives and they will usually set about finding a new partner for the following breeding season.

FEATHERED FOLKLORE

Around much of Europe, the tale of people being able to transform into swans is commonplace in folklore. Swan maidens are mythical creatures who could shapeshift from human to swan form, usually requiring a swan skin or other garment with swan feathers attached to it. In Germanic folklore, Wayland the Smith falls in love with a swan maiden, Swanhilde, the daughter of a fairy king and a human woman. Wayland rescues her when she is wounded in the air by a spear. Swanhilde marries Wayland, relinquishing her wings and the magical ring that enables her shapeshifting power. She is later kidnapped, but Wayland finds her and, though he is crippled by his enemies, fashions himself wings and helps them both escape.

In Irish mythology, the Scandinavian princess Derbforgaill is desperately in love with the hero Cú Chulainn and turns herself and her handmaid into swans to be near him more swiftly. On seeing the birds, Cú Chulainn shoots her down with his sling, unaware of her true identity until she transforms back into the princess. He saves her life by sucking the stone out of her side, but alas, then cannot marry her as you cannot marry someone whose blood you have tasted!

WHERE TO SEE: *Mute Swans can be found on almost any large body of water, particularly if there's plenty of food on offer. The Abbotsbury Swannery in Dorset has maintained its Mute Swans for centuries and allows you to walk right through a breeding colony of hundreds of birds. Bewick's Swans can be found across southern England from the Fens of East Anglia across to the Severn Estuary, with the Slimbridge and the Ouse and Nene Washes being particularly good sites. Although they are found much more widely across Britain, the eastern Fens are perhaps the best spots for Whooper Swans too, with over 10,000 now visiting the Ouse Washes alone.*

WHEN TO SEE: *Mute Swans are visible all year round but, although small numbers of Whoopers can be found nesting in the UK, they and Bewick's are winter visitors. Look for them arriving on the winds of winter and making the wetlands of Britain their home between October and March.*

Geese

Whether they're aggressively eyeing up your lakeside sandwiches or flying over in their awesome V-shaped skeins, geese are a familiar and well-known part of our birdlife. There are quite a few species, so how many geese actually are there?

There are seven found here, although five of those are only here for part of the year. Barnacle, Bean, Brent, Pink-footed and White-fronted arrive on our shores in autumn as they escape the colder temperatures of their northern breeding grounds. Their return has marked the turning of the seasons for all those who have lived alongside them for thousands of years, whilst their departure in spring tells of better times ahead.

Canada goose

Canada Geese and Greylag Geese

To many of us, there will only be two types of geese we regularly encounter, the ones found on ornamental lakes and duck ponds up and down the country.

One is the Canada Goose, with its long black neck and white chinstrap making for a striking-looking bird. They're large geese that swagger round with scarcely a care in the world, uttering their characteristic 'honk' sound as they go. As their name suggests, the Canada Goose isn't a native species to the UK, first being introduced to Britain by King Charles II and added to his waterfowl collection in St James's Park in London in the seventeenth century. In the centuries that followed, they've gone on to become one of our most successful waterfowl species given their happiness to live alongside humans.

The Greylag Goose is the other species regularly found in parks and will be recognisable to most people who've ever tried to enjoy a quiet picnic. It used to be that the only wild breeding Greylags in the UK were found in northwest Scotland and the Western Isles, but feral populations began to spring up thanks to introductions in the twentieth century, which have now seen them colonise much of Britain.

It is Greylags that are the original 'farmyard goose' and the ancestors of many domesticated breeds. Their taming was believed to have occurred at least 3,000 years ago at the hands of the Ancient Egyptians, making them one of the first animals to be domesticated by humans.

WHERE TO SEE: *Both species are very widely distributed and common in many wetlands and parks with large bodies of water.*

WHEN TO SEE: *Any time of year!*

Barnacle Geese

Arriving from the far north to winter in northern Britain, the Barnacle Goose is a small, sharp-looking species donned in blacks, whites and bluish greys. They descend from the heavens in their thousands to feed on the fields.

FISH OR FEATHER?

The Barnacle Goose shares a curious link to the oppositely named Goose Barnacle. The Goose Barnacle is also known as the goose-neck barnacle, owing to the fact that the shell is on a long stalk that attaches to the hulls of ships, piers, floating timber or anything else they can find. Occasionally, wood-carrying Goose Barnacles would wash up on the British coastline with their cirri, the apparatus they use to filter feed, poking out from the shell and resembling a downy feather. As the Barnacle Geese bred in the Arctic, it remained a mystery where the birds disappeared to or how they reproduced, so a connection was made between the barnacles and the geese.

Giraldus Cambrensis, a Welsh historian and traveller, wrote in the 1100s, 'I have frequently, with my own eyes, seen more than a thousand of these small bodies of birds, hanging down on the seashore from one piece of timber, inclosed in shells and already formed …'

As the Goose Barnacles would often wash up on wood, it was believed they originated from a special 'Goose Tree'. The archdeacon of Brecon, Gerald of Wales, wrote in 1597 how there was a tree in northern Scotland and Orkney that grew white shells upon its branches. When they reached maturity, 'Out of them grow those little living things, which falling into the water do become fowles.'

So convinced were some that Barnacle Geese were born from the shells that some Christian churches allowed them to be eaten on days that meat was forbidden. Up until the nineteenth century, in parts of Ireland, it was still considered acceptable to eat one on a Friday, a day meant for religious abstinence from meat. Sceptics date back

Barnacle goose

as far as the myth itself, however, with the Holy Roman Emperor Frederick II, a keen ornithologist himself, writing in 1241:

> *We have made prolonged research into the origin and truth of this legend and even sent special envoys to the north with order to bring back specimens of these mythical timbers for our inspection. When we examined them we did observe shell-like formations clinging to the rotten wood, but these bore no resemblance to any avian body. We therefore doubt the truth of this legend in the absence of corroborative evidence. In our opinion this superstition arose from the fact that barnacle geese breed in such remote latitudes that men in ignorance of their real nesting places invented this explanation.*

WHERE TO SEE: *Although some resident feral flocks now exist in the UK, the biggest wintering flocks are found on the Hebrides, western Ireland and the Solway Firth.*

WHEN TO SEE: *Winter is the best time to see them, when over 100,000 birds arrive from Greenland.*

Pink-footed Geese

Pink-footed Geese often first appear like a whisper on the autumnal winds, enough to make you stop in your tracks to get a better listen. Then, as the faint 'wink-wink' sound becomes clearer, their shapes begin to materialise from the sky as the skein appears. They fly in their iconic V shape to reduce drag for the birds behind them in the formation, the lead bird swapping regularly to make sure they all get a turn of easy flight. These sights and sounds of the Pink-feet heading south, filling the big blue autumn skies, are, for many, one of the defining markers that winter is coming.

The entire population of Greenland and Iceland arrive here to spend the winter, making the British Isles home to hundreds of thousands of birds and most of the world's population. Their numbers have risen spectacularly over the last few decades, leading to some seriously impressive flocks in parts of the country where they tend to concentrate. Pink-footed Geese usually spend the daylight hours feeding on farmland, consuming grain, grass and potatoes. In Norfolk, they have taken a liking to sugar beet and eat the green tops left on the ground post-harvesting.

WHERE TO SEE: *The greatest concentrations are found in Norfolk, which has around a third of the UK's winter population, followed by Lancashire and Aberdeenshire. Holkham Estate, WWT Martin Mere and the Montrose Basin are particularly good places to see them.*

WHEN TO SEE: *They begin to arrive in September and are present right through until spring before they return north.*

Pink-footed goose

Mandarin Ducks

A male Mandarin Duck is an extraordinary-looking bird. It looks like it's been crafted as either a fine art piece or some sort of toddler 'paint-by-numbers' bird. Throw every descriptor of gaudy you can at it and it still won't quite do it justice. Extravagant, resplendent, ornate – it has almost every colour you could imagine on its body somewhere and, if that wasn't enough, a crest and *sails* to boot.

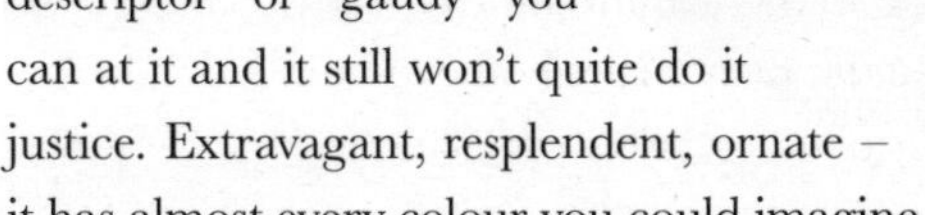

Mandarin duck

HOME FROM HOME

It's no surprise that the Mandarin was prized as an ornamental duck, worthy of any waterfowl collection. It was this that brought some from their native range in eastern Russia, China and Japan to the UK, where escapes in the 1900s led to the formation of a wild population. Numbers have continued to climb ever since, with latest estimates putting the breeding population at over 4,000 pairs. This now represents a considerable percentage of the global population.

Mandarins nest in trees close to water, making them much safer from potential predators whilst the female incubates her eggs. It may surprise you to learn that this is a strategy shared by a couple of native duck species too, but it of course poses a big question – how do the young ducklings get down?

STARS OF THE SCREEN

One of our star cameras from *Springwatch 2018,* when we were based on the Sherborne Estate in the Cotswolds, featured a female

Mandarin Duck incubating a clutch of eggs in a big old tree. The female is far more subdued than the male but no less elegant, with scalloped flanks, a stylish crest and other intricate details. We watched as she incubated her eggs diligently until they hatched and then, on 5 June, the time came for them to leave the nest.

At 7:40 in the morning, the ducklings scrambled to the entrance hole and quite literally launched themselves into the big wide world. A camera stationed at the bottom of the tree caught the moment they hit the floor with a bounce, only to watch them immediately shrug it off and follow their mother to the water. Their bodies are so lightweight at this age that, even if they can't fly, the fall isn't heavy enough to cause them any harm. It was a fantastic moment to catch on camera and one that went viral around the world when it was uploaded onto the internet!

WHERE TO SEE: *They are found across England and parts of Scotland and Wales, but they can sometimes be elusive and nervous. There are some places where they are very acclimatised to people, however, such as Endcliffe Park in Sheffield, which allows for excellent close-up views.*

WHEN TO SEE: *Mandarins can be seen all year round, although they become more secretive and difficult to find when they pair up for breeding. Winter is a good time to see them when they flock together, sometimes in quite large groups.*

Gamebirds

The gamebirds are so named because they were, or are still, widely hunted. The UK is home to six native species: Red Grouse, Black Grouse, Ptarmigan, Capercaillie, Grey Partridge and Quail. Added to this are non-native Pheasants and Red-legged Partridges, which have naturalised populations but are also reared and released into the countryside each autumn in large numbers to be shot.

Capercaillies

The biggest of all the gamebirds, the Capercaillie presides over the pinewoods of the Highlands. It is here that the males strut their stuff in forest clearings, filling the air with a bizarre sequence of pops, clicks and scratching sounds as they attempt to woo a female. Testosterone in the males peaks during lekking season and some birds have been known to be especially aggressive at this time, so much so that they will even chase and attack people who enter their territory!

Capercaillies are one of the most sexually dimorphic birds in the British Isles, with males and females not only differing in colour but also in size. Females weigh about half as much as the males and are a cryptic mix of browns, whites and blacks that help to camouflage them when incubating their ground nests. The heavyweight males can come in at up to 5kg and are coloured with much darker browns and greys, with deep metallic emerald breast feathers and a large tail, which is fanned for display. Although not quite as elaborate, it is that which has led them to be described in the past as 'Peacocks of the woods'.

PINE-FUELLED POWERHOUSE

In the winter when soft leaves, fruits and insects become scarce, Capercaillies switch to eating a lot of conifer needles. They are adept in trees, despite their big, bulky size, and are able to perch comfortably on the branches of pine trees to feed. To digest the tough needles, Capercaillies have two appendices, which grow in length during the winter and contain a diverse group of bacteria that help to break them down.

THE STORY SO FAR

The name Capercaillie is from Scottish Gaelic meaning 'the horse of the woods', seemingly a reference to the size of the bird. Given this, it is unsurprising that the Capercaillie was a valued and much sought-after gamebird. By the 1780s, the last Capercaillies had disappeared from the UK. Exactly why is still a mystery, but it is theorised to be a combination of habitat loss and deterioration, hunting and climate change, particularly the cold weather associated with the 'Little Ice Age'.

Despite earlier attempts, it took until 1837 before Capercaillies were successfully reintroduced back to Scotland. A desire to reinstate them as a gamebird led to them being brought over from Sweden and over the next few decades their population grew to around 10,000 pairs. However, since the 1960s, the Capercaillie has undergone a steep decline and the picture is not so rosy, with an estimate of only around 500 birds at the latest count.

WHERE TO SEE: *Due to the Capercaillie's critically low numbers and vulnerability to disturbance, it is not advised to go looking for them.*

WHEN TO SEE: *You increase your likelihood of a chance encounter if you visit the Highland pinewoods in autumn or winter, when the Capercaillie may be seen visiting gravel tracks to obtain the grit they need to digest the tougher food they eat at that time of year.*

Ptarmigan

Ptarmigans

As morning breaks on the steep slopes of Scottish mountains, following a night where winter's icy grip plummeted temperatures to well below freezing, a snow-covered boulder begins to stir. It rises out of the deep snow until it is sitting on the surface and it is only then that the identity becomes clear. A Ptarmigan. It has spent the night in a snow hole, dug the previous evening to shelter it from the worst of the bitter cold.

COSTUME CHANGE

Ptarmigans are compact little grouse. They change their feathers with the seasons, going between browns and greys through spring, summer and autumn to camouflage them against the bare ground, whereas in winter they don feathers of purest white to disguise them against the snow from any passing eagles.

Ptarmigans are the Ernest Shackleton of the British bird world, being better adapted to surviving freezing, exposed conditions than any other. As well as their ability to dig into the snow, they also have completely feathered legs and *eyelids* to stop as much heat from escaping as possible.

WHAT'S IN A NAME?

Their name comes from the Gaelic *tàrmachan* (croaker) but gained its current spelling when later English writers falsely believed that the word must be related to the Greek word for 'wing', *pteron* (as in pterodactyl). So, the word *tàrmachan* was erroneously changed to *Ptarmigan*, and we've been spelling it wrong ever since!

WHERE TO SEE: *Although present on a few of the Western Isles and dotted around other areas of Scotland too, by far the best numbers are to be found within the Cairngorms National Park. The ski centres at Glen Shee or on Cairngorm Mountain are regular spots for birders to look.*

WHEN TO SEE: *Notoriously tricky to spot at any time of year given their camouflage, but winter will at least bring them lower down the slopes as they seek more shelter. Winters with little snowfall make their white plumage very easy to see.*

Quails

One of the features of the gamebird family is how built for life on the ground they are. Although Pheasants, Partridges and Grouse can all fly, none of them aim to travel particularly far, mainly using flight for escaping predators or roosting off the ground.

The Quail stands alone in taking an altogether more ambitious approach when it comes to the gamebirds' use of flight. The birds that live in the British Isles throughout spring and summer will, at the onset of autumn, take to the skies under the cover of darkness and wing their way across Europe, over the Mediterranean Sea, across the Sahara and into Central Africa. All this for a bird less than 20cm long that spends its days scurrying around on the ground like a feathered mouse.

Seeing a Quail can feel like an impossible task given their size and the fact they make their homes in tall grass, meadows or farm crops. They are best detected by their sound: a loud, repetitive three-note call that's been described over the years as wet-my-feet, but-for-but or, most famously, wet-my-lips. They are not a common bird but numbers can fluctuate quite dramatically from year to year for reasons that are still a mystery.

WHERE TO SEE: *Quails can turn up anywhere in lowland Britain wherever there is plenty of arable farmland or open grassland.*

WHEN TO SEE: *Listen for their call when they return to breed in late spring and early summer. They sing mostly at dawn, dusk or even through the night.*

Quail

Grebes

The Great-crested Grebe is most well known for its party piece. On a calm water surface on an early spring morning, a pair of grebes will bond together with one of the most mesmerising courtship displays in the bird world. The dance is complex and involves a choreographed mix of calling, head shaking, mirror swimming and synchronised preening. Their most famous flourish comes as the dance reaches its crescendo, however. Most birds will split apart, diving beneath the water to find some underwater vegetation to hold in their beak.

Once both birds have their adornments, they begin to swim quickly towards each other. As they meet, they rise up on furiously paddling feet to lift their whole bodies from the water. They hold themselves there, breast to breast, flicking their heads back and forth alternately in what is known as 'the weed dance'.

The look of the birds themselves adds to the whole performance. The Great-crested is one of the biggest grebes in the world and is easily told apart from other waterfowl by its dagger-like bill. Although it does possess a couple of tufts on the top of its head, its crest is much more reminiscent of a mane. When fanned, this impressive ruff radiates out in a striking orange-and-black colour scheme set against their white face and red eye.

PLUMAGE PROTECTION

It was these fine plumes that almost caused their downfall when, in the nineteenth century, the grebes were ruthlessly hunted so their feathers could be used to decorate ladies' hats. Such was the demand for these feathers that the birds were at serious risk of extinction. Thankfully, there came a saviour in 1889 in the form of Emily Williamson who, sickened by the slaughter of birds for the fashion industry, founded what would soon become the Society for the Protection of Birds. In 1921, her campaigning efforts against this 'murderous millinery' brought about the Plumage Prohibition Act, which safeguarded the likes of grebes, egrets and many other birds from all over the world from being killed for their feathers. Her society has continued to protect and campaign for nature ever since, becoming one of the biggest nature organisations in the world – the RSPB.

MEET THE FAMILY

There are five species of grebes found in the UK. The Red-necked is a real birder's bird, only turning up in the winter to hang around off the coast in its drab non-breeding plumage. The Slavonian and Black-necked Grebes are rare nesters, only found on a handful of sites. Other than the Great-crested, it is the Little Grebe that you're far more likely to encounter.

This pocket-sized water bird is only around 25cm long and its tiny size gives it a very cute appearance. Little Grebes have had many names throughout history and whilst Dive-dop, Diedapper and Dobber have all fallen by the wayside, Dabchick is still used by many.

TOUGH TO STOMACH

All grebes catch their food underwater, whether it be fish, insects or other aquatic invertebrates. On top of this, a curious side dish for grebe chicks is feathers, which they are regularly fed by their parents as they grow. Feathers contain no nutritional value, so the reasoning is a little curious. One of science's best guesses relates to how birds digest their food. Much of the grinding process in bird digestion occurs in the gizzard before the food passes into the intestines to be absorbed. However, in predatory birds, the gizzard is poorly developed as flesh often requires less breaking down for it to be absorbed successfully. Consequently, if you're a newly

DID YOU KNOW?
All grebes fall within the family Podicipedidae, which comes from the Latin meaning 'rear-end foot' and refers to how far back the legs are positioned. This aids them in being excellent swimmers underwater but means they're so ungainly on their feet that they rarely ever set foot on land.

hatched grebe chick being fed a crustacean or bony fish, the danger of having some of the hard, undigestible bits of food pass into the intestine could spell trouble.

This is where the feathers come in, and it's suggested that the feathers accumulate in the gizzard, forming a dense mass that traps the other incoming food and allows it to be broken down for longer before it enters the more sensitive intestine!

WHERE TO SEE: *You stand a good chance of spying a Great-crested or Little Grebe on most decent-sized water bodies, particularly on wetland nature reserves. Black-necked and Slavonian Grebes only breed in small numbers in quiet areas, so it is best to leave them undisturbed. Both of these species, and the much rarer Red-necked Grebe, can be found wintering off the British coastline in winter.*

WHEN TO SEE: *The time to see Great-crested Grebes is when they are looking at their best dressed in their breeding finery. They perform their famous 'weed dance' at the very beginning of spring, in February or March, to strengthen their bond before raising a family.*

Divers

These are magical birds of still Scottish lochs and inlets, whose haunting cry penetrates through the mist and seems to sit upon the air long after the bird has finished. It is only the Red-throated and Black-throated that breed here, but both are exquisitely adorned in detailed patterning when in their summer plumage. In winter, they leave for the coast and are joined by the Great Northern Diver from, well, the north. Here they take on a drabber winter plumage and may be confused at a distance for a cormorant hunting low in the water.

LORE OF THE LOONS

Their mournful cry has embedded them deep within the folklore of their northern haunts. Divers get their alternate name of 'Loon' from the Norse *lóme*, meaning 'to moan'. Norwegians believed that their wails were those of ghosts or water spirits and worried about impending doom upon hearing them. In native North American culture, the loons play important roles in traditional stories. They serve the gods, they aid humans searching for fish and they even have the ability to restore lost sight and heal the sick and injured.

In the Faroe Isles, the Red-throated Diver was believed to escort souls to the afterlife, whilst further south, in Shetland, they were watched with a keen eye as a signal of what the conditions out to sea were like. Here they were known as the 'rain-goose' and an old rhyme tells of how a diver flying inland is a sign it's safe to brave heading out into the open ocean. If she's seen heading out to sea, however, it's time to take shelter because a storm is coming!

WHERE TO SEE: *To experience these birds at their best, when their exquisite plumage and haunting calls can be fully appreciated, visit the quiet lochs and lakes of north and west Scotland. These are the best places to find Red-throated and Black-throated Divers during the summer.*

WHEN TO SEE: *If the journey north is a bit much, then fear not, as Divers can often be found on the sea near you in the winter. Both of our breeding species and the Great Northern Diver turn up anywhere around the entire British coastline and you stand a good chance of spotting them if you train your binoculars out to sea. Beware that they will be dressed in their trickier winter plumage however!*

Storm Petrels

The appearance of Storm Petrels was a bad omen for mariners keeping a weather eye on the conditions. They were said to foretell, or even cause, bad weather, resulting in rough seas and a danger to anybody aboard a ship. Mother Carey, a supernatural figure said to be the personification of rough seas and the danger they bring, was said to work closely with Storm Petrels – giving them the old common name 'Mother Carey's chickens'. Sometimes they were said to be the souls of perished sailors, arriving in the tumultuous weather that had sent them to their watery graves, and killing them was believed to be bad luck for those with lives linked to the sea.

STORM SURFERS

The Storm Petrel's tie to turbulent seas is a logical one when we consider its biology. This tiny seabird, little bigger than a sparrow, belongs to the same order of birds as those famed ocean wanderers, the Albatross. Just like their bigger brothers, Petrels rely on the ocean winds to aid their flight and rest on the water's surface if it isn't there to aid them. As such, their swirling presence around boats out to sea would coincide with an uptake in wind speed and more dangerous conditions.

Their link to storms may explain the first half of their name – but what about the latter? The word 'petrel' was first recorded in 1602, as a corruption of 'pitteral', a reference to how Storm Petrels seem to walk on water, picking organisms from the water's surface.

WALK ON WATER

Most of their food is zooplankton, although squid, jellyfish, small fish and crustaceans will be taken too. When feeding, they hover just above the water with their wings outstretched and their legs reaching down, touching the surface. From this pitter-patter of their tiny feet arose the name 'pitteral', which was shortened to petrel. Another theory is that the name comes from a form of 'Peter', in reference to Saint Peter. In the Bible story of Jesus walking on water, Peter is initially described as being able to do the same, before beginning to sink as his faith wavers. It may be that the Petrel's light-footed tiptoeing on the waves, followed by their habit of sinking down onto the surface of the water to rest, brought to mind the actions of Saint Peter. This behaviour also gives them their scientific name – *Hydrobates pelagicus* – the water walker of the high seas.

DID YOU KNOW?
Despite their small size, the Petrel can live an astonishingly long time. The record, set in 2017, was a bird that clocked in at 38 years old!

SEA LEGS

Storm Petrels are so adapted for life on the open ocean that they're incapable of walking on land. They only return to land at night, to avoid predation from hungry gulls, and shuffle clumsily on the ground to reach the burrows that they nest in. Given their size and vulnerability, they only breed on Britain's islands and a handful of isolated headlands that mammalian predators struggle to reach.

WHERE TO SEE: *Seeing Storm Petrels is tricky, given their habit of nesting in remote areas and only coming to land at night. Boat trips can be booked in hot spots such as Shetland and Pembrokeshire to see the birds out on the ocean during the daytime.*

WHEN TO SEE: *Autumn is your best chance of spotting a Storm Petrel in the day from the shore. Find a spot on a headland that sticks out into the sea on a day when the wind is blowing strongly onshore. These conditions can push them closer to land and give you the best chance of seeing them!*

Manx Shearwaters

As night falls on Trollaval Peak on the Hebridean Isle of Rùm off the west coast of Scotland, a strange sound begins to surge. The black summer air is filled with shrieks, cackles and screams, a supernatural chorus of eerie cries emanating from the darkness. Those in the past believed the cacophony to be the sound of Trolls, with Trollaval meaning 'mountain of the trolls' in Old Norse.

The perpetrator of these calls may be far from mythical, but the Manx Shearwater is no less a wondrous beast. These are highly specialised seabirds, closely related to Storm Petrels, Fulmars and Albatrosses. With long, flat wings, they can glide with few wingbeats, skimming the surface of the water as they search for the small fish and other marine life that they feed on.

Manx shearwater

ISLAND FORTRESS

Most of the world population of Manx Shearwaters breeds in Britain and Ireland, nesting in burrows on small islands that are usually free of rats. Rats are bad news for Shearwaters, taking eggs, chicks and even adult birds when they're underground in their nests. One of the largest colonies to exist used to be found on the Isle of Man, giving the bird its name, but an accidental introduction of rats from a shipwreck in the late eighteenth century caused their extinction everywhere but the Calf of Man.

Fifty per cent of the UK population nests on the Pembrokeshire islands of Skomer, Skokholm and Middleholm, with 40 per cent on the Isle of Rùm, now home to the current biggest breeding colony. Whilst there may be rats on the Isle, the mountain slopes that the birds nest in seem to keep the Shearwaters out of harm's way.

BORN TO WANDER

One egg is laid each year, with the chick growing slowly over two months until it reaches fledging age. Only a few days after leaving the nest, the young Manx Shearwater has to begin its journey to its wintering grounds off the coast of Brazil and Argentina. Its parents will have already left, and the young bird will have to make its southward journey alone on a migration that can be over 6,000 miles.

RECORD SETTERS

Manx Shearwaters hold the record for the longest-lived bird in Britain after a bird ringed in 1957 was recaptured alive and well at its nest burrow on Bardsey Island in May 2008. This made it the grand old age of at least 50 years, 11 months and 21 days but, as it was ringed as an adult bird of breeding age in 1957, it was likely already 5 years old at the time! It's been estimated that, with the distance covered on migration and on their ocean foraging trips, a bird of this age could well have flown over 5 million miles in its lifetime!

WHAT'S IN A NAME?

The scientific name of the Manx Shearwater is *Puffinus puffinus* and must surely be one of the most confusing. In fact, the Manx Shearwater is the original Puffin, with the name and its derivatives poffin, pophyn and puffing being used for the delicacy of a cured carcass of a fat, nestling Shearwater. Over time, the word Puffin gradually began to shift to be used for the bird we know today, but the Shearwater's scientific name has remained the same.

Manx Shearwaters' habit of only coming ashore at night is to avoid predation and has earned them the name Night Bird on the Skellig Islands of Ireland. They were also known as Mackerel Cocks and Herring Hawks after their prey, whilst Wind Fairy describes their mastery on the wing. Their name 'Shearwater' is a descriptor of their flight, tilting in the air with their wingtips so close to the water it looks like they're slicing the waves.

WHERE TO SEE: *Their nocturnal habits when returning to land make them tricky to see but staying overnight on the Welsh islands of Skomer and Skokholm gives you the best possible chance. Numbers are also increasing on the Isles of Scilly and they can regularly be seen during the day from boats on marine wildlife safaris.*

WHEN TO SEE: *Manx Shearwaters return to Britain in March from their trans-Atlantic ocean wanderings. They'll stay until August before returning to the open ocean to spend the winter in the Southern Hemisphere.*

Gannets

If asked to name the most 'British' birds you can think of, I'm willing to bet Gannet wouldn't feature on many a list. However, whilst many birds found throughout Britain are also found across Europe in even greater numbers, it's our seabirds that really set our islands apart.

More than half of the entire world population of Gannets breed around the coast of the British Isles. Its scientific name *Morus bassanus* is even derived from Bass Rock, a rocky outcrop in the Firth of Forth that's home to the largest Gannet colony in the world.

UNMISTAKABLE

Gannets are impressive birds. A 2-metre wingspan makes them by far our largest seabird and their striking colour scheme is easily recognisable. They reach their full adult plumage in 3 years after starting with dirty brownish feathers, like the colour of roadside snow after it's been on the ground for days and splattered with sludge from passing vehicles. As each year passes, the white spreads until, in their first breeding season, the black has been banished to the tips of their wings and the resplendent white now dominates most of the bird's body. The only feathers on an adult bird that aren't black or white are found on the head, where the white diffuses into a soft, powdery lemon yellow. If you're able to get close to a Gannet, perhaps as they ride the air currents as they skirt the top of sea cliffs, you may be lucky enough to get fixed in the glare of their large crystal-blue eyes that seem to contain the very element of the sea itself.

DID YOU KNOW?
The Gannet was deemed to have voracious eating habits, perhaps linked to the size that the chicks achieve before fledging. 'Gannet' has been recorded since the mid-nineteenth century as a word used to describe a greedy person who eats lots of food.

STRIKE FORCE

The Gannet's body is all about getting its favourite food – fish. Evolution has perfected them into a shape that bridges the distance between sky and water as fast as possible. Watching Gannets hunt is like watching a bombing raid. After locking onto their targets with their acute vision, they fold their wings back so that they resemble a feathered cruise missile. They plummet vertically, striking the water headfirst at speeds of over 60mph. The fish don't stand a chance. It's like being hunted by a bolt of lightning.

Gannet

ON THE MENU

Given the Gannet's size and the ease of catching them at their colony, they have long been eaten as food. Chicks, known in Scottish Gaelic as *gugas*, were harvested in the summer, just before they could fly and when they were at their fattest. The guga harvest took place for at least 350 years on Bass Rock until it stopped in 1885, whilst hunting on St Kilda didn't stop until 1910. The meat was usually served roasted and was a favourite of the poor and Scottish royal banquets alike. In the Second World War, it even made its way to London restaurants, where it was served as 'Highland goose'.

A NEW CHALLENGE

In 2022, an unprecedented outbreak of avian influenza ripped through seabird colonies all around the coast of the UK. Many thousands of birds of different species were lost, but Gannets were one of the hardest hit. Given how long it takes them to reach breeding age, their ability to repopulate back to their former numbers will take some time, and the disease also reappeared in the summer of

2023. Thankfully, there is evidence that some birds have developed immunity, with a tell-tale sign being their iris colour turning from blue to an inky black.

WHERE TO SEE: *Although not breeding in Northern Ireland, large Gannet colonies can be found in Wales, England and Scotland. Grassholm off the coast of Pembrokeshire, Bempton Cliffs on the Yorkshire coast and Bass Rock in the Firth of Forth represent the largest gatherings in their respective countries.*

WHEN TO SEE: *Gannets are best experienced in their raucous breeding colonies during the summer months, but they can be seen much more widely in the winter by scanning out to sea with binoculars or a scope.*

Herons

There's almost something a bit Dickensian about Herons. Sitting hunched on long legs, looming with an air of menace, they look like a character that could have walked right out of a Victorian novel. They're an odd bunch. Some, like the Bittern, are so reclusive that most people will never see one, whilst the Egrets strut around bright white and obvious with plumes wafting wildly in the air. As we restore our wetlands and the climate warms, we're now seeing more exotic species visit our shores, adding to our cast of these charismatic characters.

Grey Herons

Look across a lake and there's often a presence lurking in the background that gets missed. Despite their size, the hunched presence of a brooding Grey Heron can often be overlooked at first glance, as they stand motionless waiting for their prey to venture within striking range.

Although fish are the main target of their dagger beak, they're just as happy taking any mammal and bird that they're capable of swallowing. In hard winters when the water freezes over, they can be seen stalking the fields with an eye for mammals. It's even been observed that herons take a liking to a particularly difficult species for most birds to catch – moles. Their patience comes in handy when it comes to watching the earth move just long enough for them to plunge their bill into an unfortunate mole's hill.

FISHY BUSINESS

It is the Grey Heron's fishing prowess, however, that was admired by the fishermen of yore. They were convinced that the heron's feet produced an oil that would attract fish towards it, and ingredients for a 1740 witch's brew to be smeared on fishing lines included the fat of a heron. So pervasive was this belief that British naturalist Frank A. Lowe, who was writing a book on the Grey Heron, decided to test the theory by dropping the extract of heron's foot into an aquarium. The fish completely ignored it.

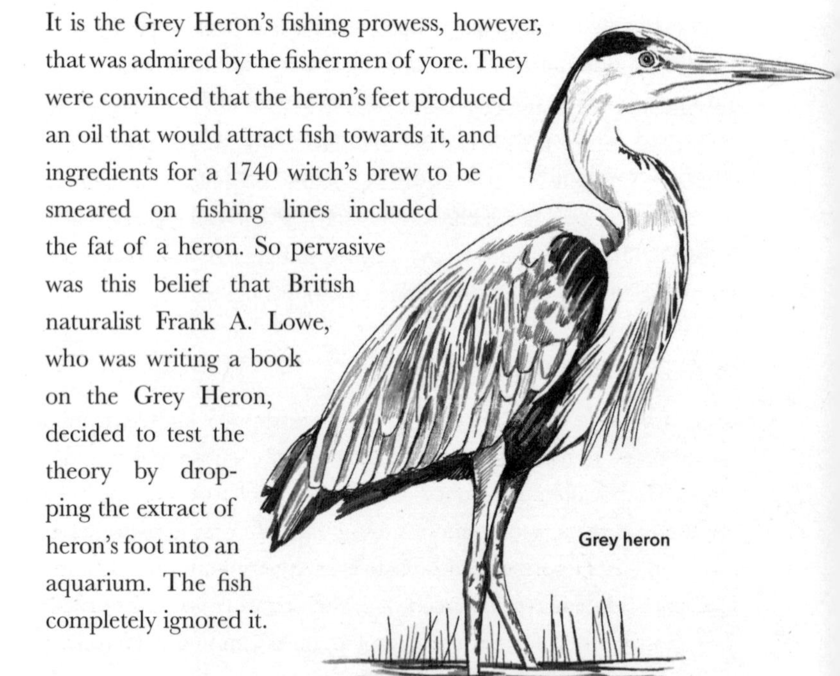

Grey heron

It was also said that the Heron's feathers glowed in the dark to allow it to fish at night. Whilst this is clearly not true, herons certainly have the capability to hunt at night and the *Winterwatch* cameras have recorded them stalking moonlit rivers in the small hours in search of a meal. It may have been the moonlight's reflection from their light, powdery feathers that sparked the myth.

BIRDY BAROMETER

Herons and their behaviour have also been linked to the weather. In Donegal, a bird flying south meant cold weather was on the way and the opposite was true if it headed north. One flying downriver meant rain was on the way, whilst upriver meant a dry spell was imminent. In Angus, an unusual superstition states that the heron waxes and wanes with the changing moon. When the moon is full, the bird is plump, when it is a crescent, the bird is lean.

FRANK!

When a solitary heron takes flight, it is often accompanied by a harsh, guttural call that can sometimes be so loud that it gives the impression they've struggled to hold it in for the hours they've spent standing in silence. This call is often written as 'frank!' and so has led to the heron being called by this human name in the past. There have been many other names too – Harn, Harnser, Harnsey, Varn, Yarn, Moll Hern, Jenny Crow, Jemmy Lang Legs and Diddleton Frank are all recorded from around the country.

HIGH-RISE LIVING

It seems an odd choice for a long, gangly bird that stands at around a metre high to nest high in trees, but that is the choice most herons make. They're often positioned at the very top of trees and the nests are used for years, getting more substantial each year as extra material is added. Heronries can be active for a very long time. On the Holkham Estate in Norfolk, records of their heronry go back to 1870. The British Trust for Ornithology has been running a Heronries Census since 1928, counting nests across the country each year. It continues to this day, making it the longest-running breeding bird survey in the world.

It's no surprise that the breeding aggregations of these huge birds attract human attention. The biggest heronry in the UK is at Northward Hill in Kent, where more than 150 pairs of herons gather in the treetops each year to make their nests. Watching a heronry is like glimpsing back into the past, giant feathered dinosaurs soaring in to land on their pterodactyl wings. As the adults come to land, they are greeted by a cacophony of prehistoric squawks and cackles from punk-crested chicks begging for food.

STARS OF THE SCREEN

The *Springwatch* cameras have been trained on breeding Grey Herons on a couple of occasions, once in mid-Wales at RSPB Ynys-hir in 2011 and again in Norfolk at Wild Ken Hill in 2022. The chicks would spend most of the time sitting rather happily in the nest together and it was only when the adults came in to regurgitate a meal that all-out war broke out in a deafening chorus of screeching.

You could have drawn up quite an impressive species list of the unfortunate animals that we saw becoming the subject of a tug of war by hungry heron chicks! As well as all sorts of fish, including some huge eels, there were plenty of birds and mammals thrown into the mix.

WHERE TO SEE: *Grey Herons are common on most large water bodies, although they'll famously visit garden fishponds too!*

WHEN TO SEE: *For the full prehistoric experience, visit a heronry in April. This will be after the chicks have hatched but before the leaves fully open on the trees and obscure them from view.*

Little Egrets

Until the 1980s, the Little Egret was a scarce migrant to Britain, only appearing as vagrants who had arrived here accidentally. Suddenly, numbers of wintering birds on the south coast began to build and, in 1996, the first pair were found breeding on Brownsea Island in Dorset. Since then, their population has gone from strength to strength, with the latest estimates showing well over 1,000 breeding pairs. They're now firmly established in the southern half of the UK and Ireland and there are breeding birds in Northern Ireland and Scotland too.

LITTLE ADAPTATIONS

Little Egrets are only around half the size of a Grey Heron, and so specialise in working the shallower water and hunting much smaller fish. They're more mobile than Grey Herons, preferring to walk the shallows and strike at the fish that flee in front of them. Little Egrets have yellow feet, so strikingly juxtaposed to the rest of their black legs that it looks like they've stood in a tin of paint. Rarely is anything without purpose in nature and, as the Egret wades through the water, it shakes these brightly coloured feet. The aim is to startle its prey, causing them to bolt and reveal themselves to the egret, which uses its lightning reflexes to make the catch.

Little egret

CLIMATE COLONISERS

Because of their need to fish in shallow water, Little Egrets have always been susceptible to cold winters, when ice can make it impossible for them to hunt. As our winters have warmed, the British climate has become far more suitable and their numbers continue to expand. In fact, the Little Egret has led the charge for more members of the heron family previously found in warmer regions to colonise Britain.

Its bigger brother, the Great White Egret, first bred in 2012 and is now an established year-round resident, whilst flocks of over 350 Cattle Egrets can now be seen in Somerset's Avalon Marshes. Night Herons, Little Bitterns and Purple Herons have also all bred in southern England since the turn of the century and with the recent news that the exotic Glossy Ibis has successfully nested in Cambridgeshire too, these wetland species serve as a reminder that what we class as a 'British' bird is changing rapidly.

WHERE TO SEE: *Little Egrets can be found in shallow estuaries or wetlands throughout all of southern Britain and Ireland. The wetlands of Somerset have proved a magnet for newly colonising species and host the biggest breeding colony of Great White Egrets.*

WHEN TO SEE: *Little Egrets are more numerous in winter when birds arrive from the Continent and roam further, turning up on smaller water courses. In recent years, an ever-growing roost of hundreds of Cattle Egrets on the Somerset Levels has provided an exotic winter spectacle.*

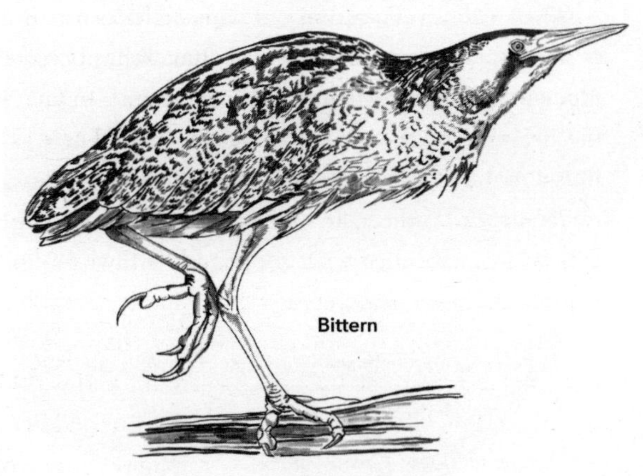

Bitterns

There is one heron left to mention. It is the most enigmatic of its family and certainly the most difficult of our regularly occurring species to see. It hides deep within the reeds, standing stock-still in amongst the phalanx of phragmites to avoid detection. Their black streaks and speckles sit upon a warm tawny plumage that gives them the ability to shapeshift and melt into their surroundings.

BASS NOTES

Bitterns live in huge reedbeds, where they are safe from disturbance, but this poses a problem when it comes to finding a mate when you're so perfectly camouflaged within your habitat. The Bittern's solution to this problem is in its call. In late winter and spring, males advertise their presence with a deep, booming fog horn-like sound. In calm conditions, the sound can carry as far as 3 miles away and for a long time it was deemed a mystery as to how a creature of its size could produce such a far-travelling sound. Theories abounded; did they make their call into the hollow reed straws to amplify it? Did they do it directly into the water or mud to give it such a low frequency? We now know that the Bittern makes the sound in its oesophagus with the aid of powerful muscles, not in the syrinx, where many other birds generate their calls from.

Their call is reflected in old names – Boomer, Bog Bumper, Bull o' the Bog – but it was a sound that disappeared during the late nineteenth century when the Bittern became extinct in the UK with the draining of the wetlands and hunting. Their rarity is reflected in Sir Arthur Conan Doyle's Sherlock Holmes classic *The Hound of the Baskervilles*. After hearing an eerie noise on Dartmoor, Dr Watson asks a local naturalist what the sound he heard might have been, to which the conversation goes:

> *'Bogs make queer noises sometimes. It's the mud settling, or the water rising, or something.'*
>
> *'No, no, that was a living voice.'*
>
> *'Well, perhaps it was. Did you ever hear a bittern booming?'*
>
> *'No, I never did.'*
>
> *'It's a very rare bird, practically extinct in England now, but all things are possible upon the moor. Yes, I should not be surprised to learn that what we have heard is the cry of the last of the bitterns.'*

Thanks to dedicated wetland restoration and reedbed management tailored to their specific preferences, the Bittern's return has been a successful one. In recent years, there have been over 200 males recorded booming on the UK's reedbeds, although being lucky enough to lay your eyes on one is a different challenge!

STARS OF THE SCREEN

One of the most privileged views that *Springwatch* has ever been able to deliver into the homes of viewers was back in 2014 when cameras broadcast the life and times of a Bittern nest from RSPB Minsmere. It was an incredible insight into the world of a very secretive, rare bird and allowed us to see some extraordinary behaviour. Although three chicks hatched, the youngest died soon after. In a classic case of nature's 'waste not, want not' policy, the female ate it and then regurgitated it to the others! The other two chicks grew well and, as we watched them turn from ugly babies to fluffy, gangly chicks, the nation took them to heart.

WHERE TO SEE: *Bitterns only nest in our largest reedbed reserves: the Avalon Marshes in Somerset, the Fens and Broads of East Anglia and Cambridgeshire and the wetland reserves of Yorkshire.*

WHEN TO SEE: *Head to Bittern hotspots from January to April to hear their evocative booming call. When chicks are in the nest, some reserves, like RSPB Old Moor in South Yorkshire, have designated spots where you can watch Bitterns flying back to the nest with food.*

Ospreys

Historically found right across the UK, the Fish Hawk would once have been a familiar sight to many. Arriving back from their African wintering grounds each year, this majestic bird would have been witnessed soaring over lakes, rivers and coastal areas, dramatically plunging below the surface and emerging from the frothing water with its fish prize.

TOOLS OF THE TRADE

Despite being classed with eagles and hawks over the years, the Osprey is its own unique bird of prey that belongs to its own family. It specialises in eating exclusively fish and has a number of adaptations to help it capture its quarry. Closable nostrils keep water from entering during dives, backwards-facing scales and sharp little spines on the talons and toes help them hold onto slippery fish, while a reversible outer toe allows them to switch their grip for a better purchase.

A TROUBLED PAST

The Osprey's tale is one that is shared with many of its bird of prey brethren. In the early 1600s, they were plentiful enough that James I attempted to train them to catch fish for human benefit along the banks of the Thames at Westminster. This didn't amount to much and even to this day, the Osprey continues to be one of the few birds of prey that seemingly cannot be tamed to a life in captivity.

During the time of the Victorians, the Osprey became ruthlessly persecuted for its eggs and skins and to stop it from taking fish stocks. By 1840, it had become extinct as a breeding bird in England and it had disappeared from its last sanctuaries in Scotland by 1916.

Osprey

ON THE UP

Ospreys were determined, however, and a pair made efforts to recolonise Scotland in 1954 around Loch Garten. After intense efforts were made to protect the nest from egg collectors, Ospreys began to spread. As numbers built up over the years, birds began to settle in northern England and Wales, with translocation projects also establishing populations on the South Coast of England and the Midlands.

FAN FAVOURITES

Ospreys are one of the most watched birds when it comes to nest cameras. They are long-lived and return to favoured nest sites year on year, meaning viewers get very attached to their Ospreys. Individuals such as Mr Rutland, EJ, KL, Flora, Telyn, Monty and CJ7 have built up fanbases across the world over the years, with avid viewers awaiting their return from West Africa with bated breath each spring. Their family dramas unfold in front of these remote cameras, with people at home enthralled as eggs are laid, chicks hatch and young birds make their first flights.

STARS OF THE SCREEN

In *Springwatch* 2023, we were lucky to be a fly on the wall in witnessing a very special pair taking their first steps to establish a new stronghold for Ospreys on the South Coast of England. Conservation efforts had meant that the previous year, this pioneering pair had produced the first chicks in southern England for 200 years, and we were able to watch as three more young Ospreys were brought into the world.

Thankfully, the Osprey's trajectory seems to be heading just one way, and it should only be a matter of time before its dramatic dives come to a water body near you.

WHERE TO SEE: *Today, there is an ever-growing list of spots to enjoy Ospreys. Loch Garten in the Cairngorms is still one of the most reliable spots to see them, as well as the Loch of the Lowes and Loch Lomond elsewhere in Scotland. In England, Kielder Forest and Rutland Water host good populations, whilst the Dyfi Osprey Project, Llyn Brenig and Glaslyn Valley are reliable Welsh sites. 2023 also saw the first successful breeding of Ospreys in Northern Ireland for over 200 years!*

WHEN TO SEE: *Ospreys disappear in the autumn to spend the colder months in West Africa. They can arrive back to our shores as early as late February, but the main arrival tends to occur in early April when they return to their favoured nest sites.*

Golden eagle

Eagles

The windswept mountains and hills of Scotland are home to perhaps the most majestic of all our birds. These are a bird without a liking for people and so it is only those with an adventurous side that are likely to be afforded the experience of looking up at a Golden Eagle effortlessly soaring on its 2-metre wingspan. As with most birds of prey, the story of the Golden Eagle is one of dramatic reduction in its range thanks to persecution. The Welsh name for Snowdonia, Eryri, translates to 'home of the eagles' and Golden Eagles were known to breed there until the mid-1800s. Records from lowland Derbyshire in the 1700s show that Golden Eagles were present in these landscapes too and the last breeding pair in the Lake District was only lost in 2004.

Now they are restricted as a regular breeding species to the Scottish Highlands and Islands, although conservation efforts in southern Scotland and Ireland should hopefully aid them in reclaiming some of their former glory.

The Golden Eagle's larger cousin, the White-tailed Eagle, had an even rougher time of it. Completely wiped out from Britain when the last bird was shot in Shetland in 1918, it took until 1968 before reintroduction attempts began to be made in Scotland. These were successful and, coupled with similar projects in Ireland, are now seeing the White-tailed Eagle return to our skies. Most recently, pairs are even beginning to breed on the south coast of England.

Even with a helping hand from humans, eagle population levels are slow to build. This is down to how long they live, with a typical lifespan for both species of around 20 years. This means young birds take their time when it comes to breeding and generally start at around 4 to 5 years old. Once breeding has begun, they build large nests on cliffs or trees that may be used for generations. The largest known Golden Eagle nest in Britain was 4.6 metres deep and had been used for 45 years!

The nest sites of these magnificent creatures became markers for humans. 'Erne', the Anglo-Saxon word for eagle, specifically seemed focused on the White-tailed, and places like Earnley in West

White-tailed eagle

Sussex still allude to its historical presence. With the positive surge in numbers in both our species, hopefully it's not long before these places become the homes of eagles once more.

Both species are apex predators in their environment, hunting a range of prey. Golden Eagles focus on a range of animals – rabbits, hares, grouse, ptarmigan and seabirds are all popular prey. The more adaptable White-tailed Eagle will also help itself to all the former, but its other common name of Sea Eagle speaks to its adeptness at catching fish from the water.

DID YOU KNOW?
Roman naturalist Pliny the Elder wrote that eagles were able to look straight into the sun without suffering any harm. It was sometimes added that they would force their chicks to do the same and any that blinked would be ejected from the nest!

TALISMANS

It's unsurprising that birds like eagles have been revered since humanity's early days. The world's oldest known jewellery are eight White-tailed Eagle talons excavated in Croatia, which show evidence of being polished and cut so that they could be strung as a necklace or bracelet. These date back an astonishing 130,000 years and show how eagles held a special place even for the Neanderthals.

On Orkney, once home to a thriving Stone Age society, a major discovery was made in 1958 when local farmer Ronald Simison chanced upon the 'Tomb of the Eagles'. Dating back 5,000 years, the tomb was a mass burial site containing thousands of human bones. In amongst them were the talons and bones of White-tailed Eagles. Exactly why the eagles were present in the tomb remains a mystery, but with eagle symbolism rife across much of the ancient world, it's easy to imagine there may have been some spiritual basis to it.

EAGLE EYES

One of the White-tailed Eagle's Gaelic names is *iolaire sùil na grèine*, 'the eagle with the sunlit eye', given its beautiful pale-yellow iris. Eagles have long been associated with the sun, given that they seem to ascend to it on warm days when they spiral ever higher on thermals until they're lost from view.

Eagle vision is so famed for its quality that it has given us the phrase 'eagle-eyed'. With eyes that weigh more than their brain and take up half the space in the skull, their vision is estimated to be at least four times stronger than that of an average human. Whilst humans have 20/20 vision, eagles have 20/5 vision, meaning they can see things from 20 feet away that we can only see from 5. They're said to be able to spot a rabbit from 2 miles and their eyes are positioned in a way to give them a 340-degree field of view.

STARS OF THE SCREEN

Springwatch has a special relationship with one Golden Eagle in particular. Throughout the 2016 series, we watched a pair in Scotland rear a single female chick. In the summer of that year, Chris Packham joined conservationist Dave Anderson to fit a satellite tracker on the bird so that scientists could follow where she roamed once she'd fledged the nest. In *Autumnwatch* of that year, the audience helped to name her Freya and we've followed her fortunes ever since. Exciting news came in 2023 when, seven years after we first met her, we got news that Freya had successfully reared her first chick in the mountains of Argyll!

WHERE TO SEE: *For a one-stop shop to stand the best chance of seeing both eagle species, it's hard to beat the Isle of Mull. Home to one of the highest densities of breeding Golden Eagles in Europe and around 20 pairs of White-tailed Eagles, there's a reason they call it Eagle Island!*

WHEN TO SEE: *Eagles can be seen at any time of year as adult birds will stick around their territories throughout winter too. Boat tours from Scottish islands offering the chance to see the White-tailed Eagles up close generally run from spring through to autumn.*

Red Kites

The Red Kite is one of Britain's most successful conservation stories.

They used to be an incredibly common bird, so abundant hundreds of years ago that they could be found in towns, scavenging the waste left in streets. Shakespeare once even referred to London as 'the city of kites and crows'. They were valued for their scavenging work in helping to keep streets clean and a Royal Decree meant that killing a kite could result in capital punishment.

POPULATION CRASH

By the sixteenth century, attitudes had completely changed and the Kite, like many other birds of prey, found itself on the list of troublesome species that competed with humans by hunting game animals that we wished to hunt. The fact that it is mostly a scavenger didn't save the fortunes of the kite and, as it became rarer, it became the focus of collectors wanting their skins and eggs.

Red kite

By 1900, they could only be found in the undisturbed valleys of mid-Wales, where the remote old oak woodlands had kept them safe from relentless persecution. Their existence hung on a knife edge, with DNA analysis discovering that the entire Welsh population was descended from just one female bird.

A TRIUMPHANT RETURN

Despite protection, the Wales population was slow to expand and so the decision was made to take young birds from Continental Europe to release in Scotland, England and Northern Ireland. The plan was a huge success, with numbers of these magnificent birds continuing to rise ever since.

Between 1995–2020, their population increased by over 1,900 per cent and they can be found in impressive numbers in some areas. The famous feeding station at Gigrin Farm in mid-Wales can attract hundreds of Kites in the air at the same time. Cloaked in russet and rufous feathers and with a wingspan just shy of 2 metres, Red Kites wheeling through the sky are a marvellous sight to behold. It's their tail that gives them away as Kites, with no other raptor possessing the long, forked shape, which constantly twists and turns to maintain their direction. The Kite's mastery of the air is so great that it is this bird that lends its name to the toy that floats upon the wind.

STARS OF THE SCREEN

On *Springwatch* we've been able to peek into the private lives of two nests on recent occasions, in Wales in 2020 and Wild Ken Hill in Norfolk in 2021 and 2022. In both nests, one of the strangest things witnessed was the unusual items being brought in by the parents. As well as prey to feed their hungry chicks, they would litter the nest with clothes, gloves and any other rags they seemed able to get their talons on.

As strange as it may seem, this is quite common behaviour for Red Kites, who, as well as lining their nest with sheep's wool, have been recorded stealing hats, socks and underwear. Their kleptomaniac tendencies have been known about for hundreds of years, with

Shakespeare even warning in *The Winter's Tale*, 'When the kite builds, look to lesser linen.'

WHERE TO SEE: *Kites can be enjoyed over much of the UK now thanks to their successful reintroduction. Mid-Wales remains a stronghold and feeding stations like Gigrin Farm offer an unforgettable spectacle. The highest densities elsewhere are found close to areas where they were reintroduced, such as the Chilterns, West Yorkshire, Northamptonshire in England, County Down in Northern Ireland and the Black Isle and Dumfries and Galloway in Scotland.*

WHEN TO SEE: *Red Kites can be seen throughout the year. They will roost communally in the winter, sometimes in groups of 50 or more birds! Look for large numbers drifting towards woodland as the skies darken during late afternoons.*

Buzzards

It's often their call that alerts you to the presence of a Buzzard. It's an unexpected sound for a bird of prey: comparatively soft, a plaintive mewing that turns your head skyward to see them lofting higher into bright-blue skies on their outstretched wings. If not glimpsed on their ascent into the clouds, then the Buzzard is often found perched on fence posts by the roadside, looking stoic and brooding.

FLYING HIGH

Buzzards hold the title of the UK's most common bird of prey after enjoying a population surge when a reduction in persecution and the banning of pesticides that caused their decline meant they could begin to rebound in the late 1960s. By 2000, their recolonisation of the UK was complete when they were confirmed to have nested in every UK county. As a species that prefers open ground for hunting, they do well in Britain's farmland landscape.

STARS OF THE SCREEN

Buzzards have a hugely varied diet that consists of more or less any animal they're able to fit in their talons. In *Springwatch* nests from years gone by, we've watched as all manner of unfortunate creatures have been delivered to the mouths of hungry chicks. Mammals are a favourite – voles, mice, rats, shrews, rabbits, hares, even moles. Birds are also eagerly consumed, but the most exotic menu was to be found at RSPB Arne in *Springwatch* 2023. This reserve on the Dorset coast, with its mix of woodland and heathland, is home to all of the UK's native species of reptile. Whilst seeing these reptiles proved to be a difficult task for our camera operators, our Buzzard family were having no trouble finding these secretive animals. They were regularly caught bringing in sand lizards and some very large grass snakes to the nest for the chicks to eat. On one occasion, the chick was even brought a still-live adder! Britain's only venomous snake is a fearsome foe for an adult bird to wrestle with, let alone a young chick, but it was made short work of by a hungry Buzzard as we watched its scales slither down into its belly.

SURPRISING SNACKS

Another favourite food of the Buzzard is a more surprising one for a bird of prey its size. You'll often see them sitting out in the open fields after they've been ploughed or following a bout of heavy rain. The Buzzards, being ever the opportunists, are looking for any worms that may be hanging around at the surface, which they'll snaffle with the same eagerness as a meatier prey item.

A COAT OF MANY COLOURS

As well as a varied diet, Buzzards also have incredibly varied plumage and come in a wide range of colours. Whilst most individuals are a warm brown colour with a light necklace across their chests, they can be anything from an almost black colour to a brilliant white over most of the body. No wonder the French call them 'La Buse variable'.

WHERE TO SEE: *Look for Buzzards anywhere with open land and scattered trees. They can often be seen from cars as they sit on fence posts or in fields by roadsides.*

WHEN TO SEE: *Although they're around at all times of year, they are most noticeable in late summer, when young birds take wing and circle into the sky following their parents, often making quite the racket as they do!*

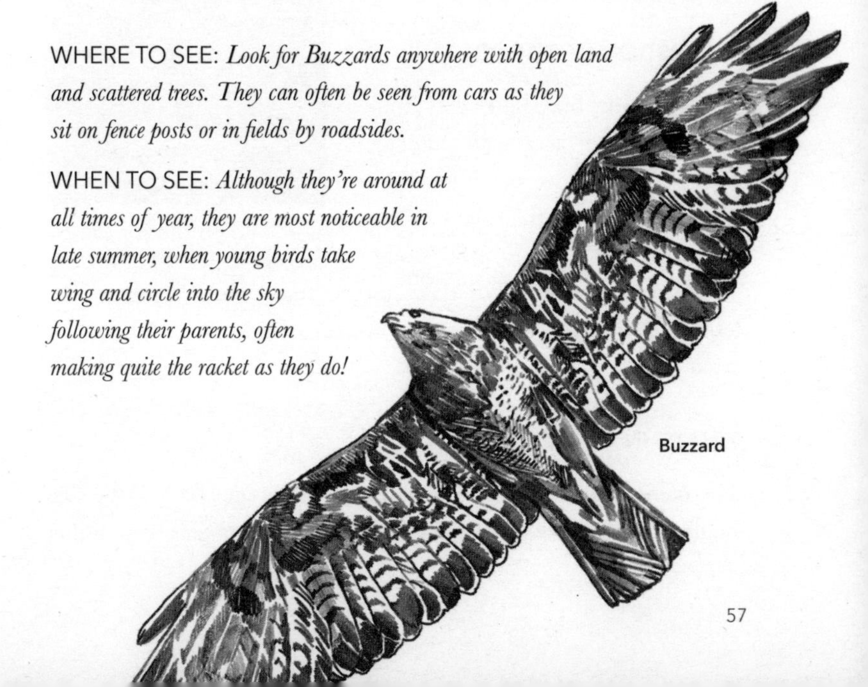

Buzzard

Sparrowhawks

Feed the birds in your garden long enough and it's likely you'll get a visit from a Sparrowhawk. It's how many of us come to see them – a flash out of the corner of your eye through the window and suddenly there it is, sat on the lawn hunched over an ex-songbird. You can see why they're not always the most welcome garden visitors.

But, like all predators, they play an important role in the food chain. As the numbers of many of our favourite garden birds have gone up in the last few decades, so have the number of garden Sparrowhawks. Their population rose steeply between the 1970s and 1990s, although recent evidence has shown a 25 per cent decline since 1995.

DID YOU KNOW?
As with many birds of prey, the size difference between males and females is large, with females reaching up to 25 per cent bigger. This allows them to split their hunting strategy, with the male focusing on smaller prey than his partner. By not targeting too many of the same birds within their territory, it means there's plenty of food to go round for both sexes.

A QUESTION OF TIMING

Just as birds like Blue Tits time their breeding season to coincide with an abundance of caterpillars, Sparrowhawks lay their eggs so that they will hatch just as lots of young songbirds are fledging the nest. This gives them plenty of naïve targets that haven't quite got the hang of flying yet, making them much easier to catch than wilier adult birds.

WHAT'S IN A NAME?

The name Sparrowhawk is old and traces its origins back to the Old English word *spearhafoc*, meaning 'a hawk that hunts sparrows'. Other names given to the bird in the past include Hedge Hawk, for its habit of hugging hedgerows when hunting, and Blue Hawk, for the colour of the male's back.

STARS OF THE SCREEN

The Sparrowhawk holds a special place in the heart of Chris Packham. For *Autumnwatch* 2015, he made a pilgrimage to see a particular male Sparrowhawk that left a lasting impression. This was Mad Max, a bird of a beautiful gun-metal blue and rich salmon-orange colour, which only mature birds obtain. He was at least 6 years old, a rare age for a bird whose average lifespan is only 4. It was the best bird Chris had ever seen and one he dubbed the finest individual bird to exist on planet Earth.

WHERE TO SEE: *Sparrowhawks are not a bird that's particularly easy to go out and look for, so you're better off letting them come to you. Spend enough time watching busy bird feeders and a Sparrowhawk is likely to pay a visit!*

WHEN TO SEE: *Sparrowhawks can be seen all year but increase their hunting activity in early summer when they have hungry mouths to feed back at the nest.*

Sparrowhawk

Goshawk

Goshawks

Goshawks are to Sparrowhawks what a fighter jet is to a paper aeroplane. They might seem superficially similar, but their auras couldn't be more different.

FEARSOME FOES

Female Goshawks can be five times larger than their Sparrowhawk counterparts and have an attitude to match. These avian terminators have just about everything on the menu. Birds the size of Goldcrests to Capercaillies have been recorded as prey, as have mammals from shrews to hares. They will regularly hunt other birds of prey too – one study (by S.J Petty et al.) looked at their diet in northern England and identified six species of raptor on the menu!

IF LOOKS COULD KILL

In adult plumage, they are clad in leaden bluish-grey on their backs and wings, with fine monochrome barring down their powerful chest and strong legs. In their earlier life, they possess an eye of pale yellow, which deepens to a molten amber as they age, sometimes even to deep, furious red. The eye colour of Sparrowhawks changes as they age too, and it's believed to be an adaptation to communicate their maturity and suitability as a potential mate.

A HELPING HAND

Goshawks are rarely seen, with these phantoms of the forest sticking to larger blocks of woodland than Sparrowhawks. Despite their secrecy, they were still ruthlessly persecuted to extinction by the end of the nineteenth century. Their return to Britain owes its thanks to falconers and hawk-keepers, who were still keeping them in captivity for hunting. Birds were deliberately released into the wild, in the hopes of creating a population where a small number of young birds could be taken back into falconry for training.

WHAT'S IN A NAME?

The Goshawk's scientific name is *Accipiter gentilis*, with the latter word being Latin for 'noble'. In the Middle Ages, it was only nobility that were allowed to fly this bird for the sport of falconry – using its formidable power and speed for hunting all manner of quarry. Their common name 'Goshawk' is from the Old English 'goose-hawk', and it was one of the few birds of prey powerful enough to be used to hunt geese.

WHERE TO SEE: *Despite an ongoing resurgence in population, their shy nature makes them incredibly difficult to catch a glimpse of. Strongholds include the New Forest, Norfolk's Breckland, Kielder Forest in Northumberland and the forests of north and mid-Wales.*

WHEN TO SEE: *Goshawks only ever make themselves showy once a year in late winter and early spring, when they perform their sky-dancing courtship high over patches of woodland to strengthen their pair bond before the breeding season. A vantage point over a good block of woodland in a Goshawk hotspot is your most dependable chance of seeing one of these secretive assassins.*

Kestrels

Look above the roadside verge and see, suspended from the sky as though on invisible wires, a true master of the air. With her head held perfectly still so as not to break her gaze on the target below, the Kestrel responds to the changes in the wind with deft flicks of her wings and tail. She is always turned into the wind, expertly shaping it with wingbeats that allow her to remain hanging as she stays focused on her prey below. If the wind begins to quiet, she will beat her wings with more force and regularity, generating the updraught she needs to stay airborne in perfect position. If she feels the chance is right, then she will fold her wings and drop from her lofty position onto the unsuspecting vole below, emerging from a tussock with it clasped between her talons.

The Kestrel's hovering skill has earned it many a name throughout time – Windhover, Hoverhawk, Wind Fanner, Wind Cuffer and *cudyll y gwynt* in Welsh – Wind Hawk. In Orkney, they call it the Mouse Falcon or Mouse Hawk, a reflection of its favourite rodent prey. Where available, it mainly prefers to target voles, but it is an adaptable little raptor that will turn its attentions to most small animals. On *Springwatch*, we've watched the Kestrels living adjacent to the open heaths of RSPB Arne in Dorset take a particular liking to sand lizards. It was the time of year when female lizards ventured into patches of exposed sand to dig their nesting burrows, which proved too good an opportunity for the Kestrels to miss.

The name Kestrel came to us from the Normans. It is derived from the Old French *crécelle*, meaning 'rattle', and may reflect their repetitive 'kee-kee-kee' cry. Another old name for them is Stannel, from the Old English *stangale*, meaning 'stone-yeller', and is derived from their habit of using rocky outcrops to watch from and nest in when in parts of the country with few trees.

Kestrel

WHERE TO SEE: *Although numbers have declined, Kestrels are still most easily seen along roadside verges as they hunt for rodents. Any open country with scattered trees or buildings is also suitable for this little raptor.*

WHEN TO SEE: *Kestrels are resident all year round but to see them at their best, look for them when a stiff breeze blows and be treated to the best of their aerial abilities.*

Peregrine Falcons

The Peregrine Falcon is one of those animals everyone knows about. Being the fastest animal alive has earned it a place in every animal-obsessed child's brain and it's a regular feature on TV shows, Top Trumps cards and animal encyclopaedias. But just how fast do you have to be to earn yourself the reputation that the Peregrine has?

RULER OF THE SKIES

In level flight, the Peregrine is nothing to write home about. It's not even as fast as its main prey, the humble pigeon. The Peregrine's secret weapon only comes into play when it is positioned above its target, with its eyes locked in preparation to make the kill. A few deep wingbeats emanating from its powerful chest muscles propel the bird downwards at speed. From here, it folds its wings back, creating a teardrop shape that hurtles steeply from the sky at an ever-growing speed towards the target below.

This is known as a 'stoop' and, at maximum speed, the falcon can be travelling at over 200 miles per hour, bridging the distance to their prey in a heartbeat. As they plunge through the air, small bony structures in their nostrils act as baffles to divert the rushing wind and reduce the air pressure entering into their respiratory system. Their nictitating membrane, a see-through third eyelid, is pulled over the eye to keep their eyes moist and free of debris.

On impact, they strike their target with their talons, often hitting their wings to reduce the risk of damage to themselves in the collision. The unfortunate bird plummets to Earth like a stone and the Peregrine will wheel around to follow it to the ground and retrieve its meal.

The Peregrine is a bird-hunting specialist and will take an extraordinary range of sizes when it comes to its prey. In North America, they've been recorded eating Hummingbirds as small as 3g in weight, whilst their stoop technique means they're able to hunt much larger birds than themselves, up to the size of swans and cranes.

Peregrine falcon

FALCONER'S FRIEND

The Peregrine's hunting prowess has long been admired and it's believed that their skills were harnessed for falconry by the nomads of Central Asia over 3,000 years ago. By the medieval period, they were held in high esteem in Europe and were widely being used to hunt game. They were placed just under the Gyrfalcon, a larger species found in colder northern countries, when it came to the bird of prey hierarchy, and so were only permitted to be flown by nobility.

Their name comes from the Latin *peregrinus*, meaning 'one from abroad', which was used to refer to travellers or pilgrims. Peregrines wander widely outside the breeding season and it's then when they would most likely have been caught for falconry. Unlike hawks, which nested in trees, the nests of falcons were often in inaccessible places, like sea cliffs, which made taking chicks from nests difficult.

PEREGRINE PERIL

This feathered thunderbolt suffered a change in perception as time moved on. With the invention of hunting rifles and the rise of game shooting, birds of prey weren't needed to provide food for the table and the sport of falconry declined in popularity. Suddenly, wild Peregrines were competitors to the quarry that humans wished to take for themselves and they were ruthlessly persecuted through the years. An indication of this comes from records on a Scottish estate at Glengarry. During a 3-year period between 1837 and 1840, 98 Peregrines were killed as part of a total of 1,372 raptors that were removed from the estate's lands.

When the Second World War arrived, the Secretary of State introduced the Destruction of Peregrine Falcons Order to protect carrier pigeons taking vital messages to and from the frontline. Many nests were destroyed and over 600 falcons were shot, mostly along the south coast, over the years the war took place.

However, the most serious threat to their population came later in the 1950s when the use of pesticides like DDT brought their numbers down to the lowest levels ever seen. The poison built up in the falcons' bodies as they ate birds that had themselves eaten insects and plants containing the chemicals. The result of this toxic build-up was fatal doses for the falcons, and those that didn't die suffered reduced fertility and laid eggs with shells so thin that they broke under the parents' weight.

CITY SLICKERS

Thankfully, Peregrine populations are now in a much healthier state, with around 1,750 pairs estimated. Part of their success has come from how enthusiastically they've embraced city living. Although there's a record of a pair nesting on Salisbury Cathedral in the 1860s, it's really in the last few decades that they've taken up high-rise living in the urban sprawl. From cathedrals to cooling towers, our manmade structures offer them the perfect substitutes for their natural cliff nesting spots and the abundance of pigeon prey means they never need to search hard for food. They're now found in the majority of UK cities, with London alone hosting as many as 30 pairs.

STARS OF THE SCREEN

In 2017, *Springwatch* paired up with Salisbury Cathedral to follow the story of their nesting pair of Peregrines. Of five eggs, only one chick hatched, but it grew strongly on a varied diet, including Kingfishers and Woodpeckers. A twist in the tale came when a nest in Shropshire became in dire need of assistance when both parent birds were suspiciously found dead. Toxicology reports proved that the birds had been poisoned and the decision was made to save the three chicks by fostering them in three separate wild nests. The nest at Salisbury was chosen as its single chick meant there would be plenty of food to go round still.

The new chick was placed in the nest box and, even though this was a tried-and-tested procedure, there was an anxious wait to see if the adult female would accept the new chick and begin to feed it. There needn't have been any worry, however, and it was no time at all before the chick was being fed and huddling up to its new nest mate. The story had a happy ending, with both chicks fledging successfully.

WHERE TO SEE: *Peregrines are most often seen in cities, where watch points are regularly set up to allow people the chance to view them during the breeding season.*

WHEN TO SEE: *Although they are most reliably seen in spring, for a chance to witness a spectacular hunt, search out Starling murmurations or large flocks of wading birds in the winter.*

Merlins and Hobbies

In addition to the Peregrine and Kestrel, there are two other falcons that regularly breed around the British Isles. To find our smallest, you must venture to the uplands where they nest upon the open moorlands, chasing down Pipits and Larks at lightning speed. These are the Merlins, a tenacious creature not much larger than a Blackbird. When conditions on the moors get tough in the winter, Merlins will often descend to the coast, following their small bird prey down to saltmarshes.

The word 'Falcon' is derived from the Latin *falx/falcis,* meaning 'sickle', referring to either the shape of its talons, beak or their scythe-like silhouette as they cruise through the sky. There is no finer example of this elegant shape that can be seen in our final falcon – the Hobby.

If falcons are the sports cars of the bird world, then the Hobby is the sports car of the falcon world. It is sleek and refined, with a dark slate-grey back and streaked front contrasting with burnt red feathering around the legs and undertail. The Hobby is our only migratory falcon and returns from Africa in spring, following the Swallows and House Martins that they like to hunt.

To be able to catch such agile birds on the wing gives you an insight into their aerial ability. Spending the summer months in our isles allows them to hunt another airborne ace too, as they can be seen spectacularly stooping over wetlands, skimming the water's surface as they snatch dragonflies in breathless blurs of sickle wings.

STARS OF THE SCREEN

Over the summer of 2014, the *Watches*' cameras were trained on a nest of Hobbies in Gloucestershire, which was brought to viewers' screens in *Autumnwatch* of that year. The three chicks were being well fed with dragonflies, Swallows and Martins when one of their neighbours took an unusual interest in the chicks. A female Woodpigeon made repeated visits to the nest, brooding the young Hobbies and even feeding them the regurgitated pigeon milk usually reserved for their own offspring. Perhaps the Woodpigeon had recently lost her own nest and the maternal instinct to care for chicks was strong. The female Hobby was quick to defend her chicks after she'd spotted the intruder and, after a few quick blows, the Woodpigeon was sent packing.

WHERE TO SEE: *Both the Hobby and the Merlin are tricky birds to find when breeding. Merlins are most numerous on the moorlands of the Scottish islands, parts of the Highlands, the North Pennines and in northwest Ireland, while Hobbies have their greatest abundance in the open farmland of the southeast.*

WHEN TO SEE: *The Hobby can only be seen between April and September and the best chance to see them in action is high summer, feeding on dragonflies over wetlands. Merlins are often easier to spot in winter, when they arrive at lowland coastal regions.*

Cranes

Hundreds of years ago, if you were to walk across almost any of Britain's wetlands at daybreak, you would have been greeted by a sound mostly unfamiliar to the modern British naturalist's ear.

Punching through the bubbling background soundscape of a waking wetland comes a raucous, joyous bugling. This is a sound that can carry more than 3 miles, so catching a glimpse of the creature responsible is not guaranteed. Luckily, this is the tallest bird found in these isles, standing at over a metre high, and it pairs this far-carrying trumpeting with an elaborate dance of leaps, bows and pirouettes.

This is the dance of the cranes, and the sight of these birds performing their exuberant courtship display is one that was lost from Britain for a period of 400 years.

CHANGING FORTUNES

Cranes used to be a common bird in Britain, with our vast historic wetlands providing them with the perfect home. There are hundreds of place names in Britain that feature the Anglo-Saxon name *cran* or *cron* – like Cranleigh, Cranwell and Cranmere. Not only were they numerous enough to give their names to settlements up and down the country, but they were also harvested for food and hunted for sport.

Falconry was a big passion of royals and noblemen in the Middle Ages, and King John is recorded as keeping large, powerful Gyrfalcons in the twelfth century for the purpose of hunting cranes. They were also eaten regularly, with perhaps the most decadent record being Henry III's 1251 Christmas banquet in York, where 115 cranes were reported to be served to the guests.

Hunting played its part in their decline, but it was the draining of the wetlands into ever smaller pockets that really made it hard for them to survive. They disappeared from the fenlands of East Anglia in around 1600 and wouldn't call Britain home again for almost half a millennium.

It wasn't until 1979 that a pair of cranes appeared in the Norfolk Broads and, two years later, they fledged the first chick. In the decades since, a lot of dedicated work in restoring wetland habitats and a reintroduction project to boost numbers have seen just over 70 pairs become re-established.

FASCINATING FOLKLORE

Cranes have long been associated with numerous superstitions. Unlike in Japanese and Chinese cultures, where Cranes are used to symbolise loyalty, good fortune, longevity and wisdom, Cranes in European cultures were associated with more unusual beliefs.

It was written in Homer's *Iliad* that the Pygmy tribes of Greek mythology were in constant battle with Cranes, who would migrate to spend the winter in their homeland. Aristotle, writing in his *History of Animals*, proclaimed that the story was far from a tall tale, stating:

> *These birds [the Cranes] migrate from the steppes of Scythia to the marshlands south of Egypt where the Nile has its source. And it is here, by the way, that they are said to fight with the pygmies; and the story is not fabulous, but there is in reality a race of dwarfish men, and the horses are little in proportion, and the men live in caves underground.*

The Celts associated birds of marsh and fenland with the supernatural, owing to the fact that they dwell in the misty 'between places' that are neither land nor water. Lugh, a figure in Irish mythology, used *corrghuineacht* (crane prayer) before battle to rally his troops and curse his enemies. Crane prayer involves chanting verses whilst standing on one leg, with one eye closed and an arm outstretched. Standing on one foot was meant to allow the spell caster to walk between worlds, just as cranes were able to visit the realms of earth, water and sky.

WHERE TO SEE: *The easiest place to see them is the Wildfowl and Wetlands Trust's Slimbridge reserve in Gloucestershire, where reintroduced Cranes can be watched with their chicks from bird hides during spring.*

WHEN TO SEE: *Other than special cases like Slimbridge, it is better to avoid searching for Cranes during the breeding season, when they are sensitive to disturbance. In winter, they gather in great numbers on the watery plains in the east of England, with numbers sometimes reaching 80 birds! Hickling Broad in Norfolk and Nene Washes in Cambridgeshire are great sites for this winter spectacle.*

Waders

Waders are a captivating group of birds, whether it is the Knot, amassing in their flocks over a hundred thousand strong and shimmering in the late summer light on the Wash Estuary, or the Snipe, taking to the skies from the rushes where it hides to perform its unearthly drumming sound above the misty bog. Perhaps it is the Avocets, so elegant and poised in their monochrome finery, sifting the water gracefully with their upturned bills, or the Dotterel, incubating their eggs with steadfast determination on the mountain plateaus. Whatever it is, the variety of wading birds we have in the UK is what makes them such a fascinating and iconic group of birds we are lucky to share our island with.

Curlew

Curlews

There is perhaps no more evocative cry in the British Isles than the rising, bubbling cry of the Curlew. It at once manages to carry a haunting, sombre tone that lingers in the air, whilst also delivering a joyful ecstasy of notes. The quality of the Curlew's cry is amplified by the landscapes in which we get to experience it. They prefer open lands, estuaries, moorland and farmland, where their calls carry and emerge ever more mysteriously through the mists.

WHAT'S IN A NAME?

It is its sound that has often been the Curlew's defining feature, despite the fact that it is our largest wader that elegantly stalks the ground, probing deep with its massive curved bill. The name 'Curlew' is a humanised corruption of its unearthly sound, with other attempts like Calloo, Courlie, Awp and Whaup all attempting to capture its essence. The shape of its bill gives the bird its scientific name *Numenius arquata*, with the former word derived from words for 'new moon' and the latter coming from the Latin for 'bow-shaped'.

ILL OMENS

Unsurprisingly, its ethereal qualities have given rise to the belief that the Curlew's cry is bad news. Any travellers should stay at home and abandon their plans if they hear its call, whilst fishermen of yore believed their call signalled an impending disaster at sea. Curlews were sometimes cast as the Seven Whistlers, ghostly birds heard at night but never seen and always a foreteller of death or ill fortune. Their wading bird cousins the Whimbrel and Golden Plover were also linked to this superstition, given their night-time flights and eerie calls.

FADING FAST

Given their size, Curlews were hunted regularly throughout the past and were one of the many species featured at the coronation feast of King Henry VI. Despite Curlew no longer being on the menu, their population finds itself at a critically low ebb. They've suffered long-term decline throughout the UK over the last few decades, enough to see them moved onto the red list and highlighted as one of the birds most in need of urgent conservation intervention.

WHERE TO SEE: *Upland regions of the UK now harbour some of the strongest breeding populations of these birds and their bubbling calls can be heard over the moors and farmland of the Yorkshire Dales, Peak District, Cairngorms and the Elan Valley, to name a few. In the winter, they can be found regularly in many coastal areas of Britain.*

WHEN TO SEE: *Winter is often the easiest time to see them, when large numbers gather at coastal areas to feed on exposed mud in tidal zones. In spring, they move to breeding areas and can be harder to find as they settle in their pairs.*

Lapwings

Rising above the fields, with their butterfly-like wings beating the air in deep, lolloping strokes, is where we find one of the UK's most beloved wading birds. Lapwings are known to many people as Peewits – an onomatopoeic name that draws on their double-note call. The name 'Lapwing' is said to come from the Old English *Hleapewince,* meaning 'a leap with a waver in it', in reference to their flight pattern. They like open ground to nest on but aren't particularly fussy about what form that takes. Grazed grassland, arable fields, moorland and salt marshes are all fair game for the Lapwing, as long as they are able to raise their chicks without much disturbance.

SNEAKY TRICKS

Like almost all waders, Lapwings nest on the ground, usually laying four incredibly camouflaged eggs that are disguised against the ground when they're not being incubated. A ground predator that veers too close to a Lapwing nest can expect to be dive-bombed by a screeching parent, whilst any avian intruder is chased off the premises. If an animal attacker gets too close to a sitting female, then she will employ her devious 'broken wing display', hopping off her eggs and walking away drooping her wing. This gives the impression that she's injured, in an attempt to lead the predator away from her precious eggs and towards her. It is only when she has led it on her merry dance and she is safely away from her nest that she will reveal it was all a ruse and take to the sky.

The Lapwing's sneaky behaviour has been known for centuries. The collective noun for them is a 'deceit' and Chaucer described them in the 1300s as 'The falsy lapwynge, ful of trecherye'. Their relentless mobbing behaviour was said to have given away the position of Presbyterian rebels who were crossing a Scottish marshland in 1685, which led to their capture, deportation and execution.

PRECOCIOUS PLOVERS

Lapwing chicks are precocial, meaning that they're able to walk, feed themselves and leave the nest as soon as they are hatched. It used to be said that Lapwing chicks were so eager to leave the nest that they would run around with the shells still on their heads, even leading to descriptions of brash people as 'lapwings with a shell upon their heads'. Although the chicks can leave the nest instantly, there are normally a few hours after hatching when they are brooded to keep warm. Lapwing eggs were often collected for food, so it is likely that the fleeing chicks were scattered by egg hunters rather than just being over eager to explore the world!

WHERE TO SEE: *Lapwings can be found in good numbers in the breeding season in many of our national parks, where they will regularly nest in land grazed by sheep or cattle. In winter, look for flocks at wetland reserves where they like to hang out on islands in open habitats around shallow lakes or scrapes.*

WHEN TO SEE: *Lapwings offer a treat for the eyes and ears all year round. In winter, they gather in large groups on wetlands or flooded farm fields, swirling in their shimmering flocks when they take to the air. In spring, look for the male carrying out his display flights as he does high-energy laps of his territory accompanied by his electronic courtship song.*

Lapwing

Oystercatchers

The Oystercatcher is a striking, instantly recognisable bird. Its black-and-white colours earned it the name Sea Pie and other such variants as Sea Piet, Sea Nanpie, Mere Pie, Sea Pyot and Pienet, which used to be much more commonly used than Oystercatcher.

SEAFOOD DIET

Oystercatcher is perhaps not the best name for us to have settled on, with oysters not forming a regular part of their diet. Old names like Mussel Cracker or Mussel Picker are far more accurate, as the bird uses its fierce orange bill to smash and prise open the shell to get at the body of the well-defended shellfish.

Oystercatchers are resourceful birds and able to make use of a range of foods. As well as smashing the living daylights out of limpets and mussels at low tide, they're also just as happy probing into mud for worms as they are sand for marine bivalves. This adaptability has meant they saw an increase along inland waterways in the seventies and eighties and can be found breeding far away from the coast.

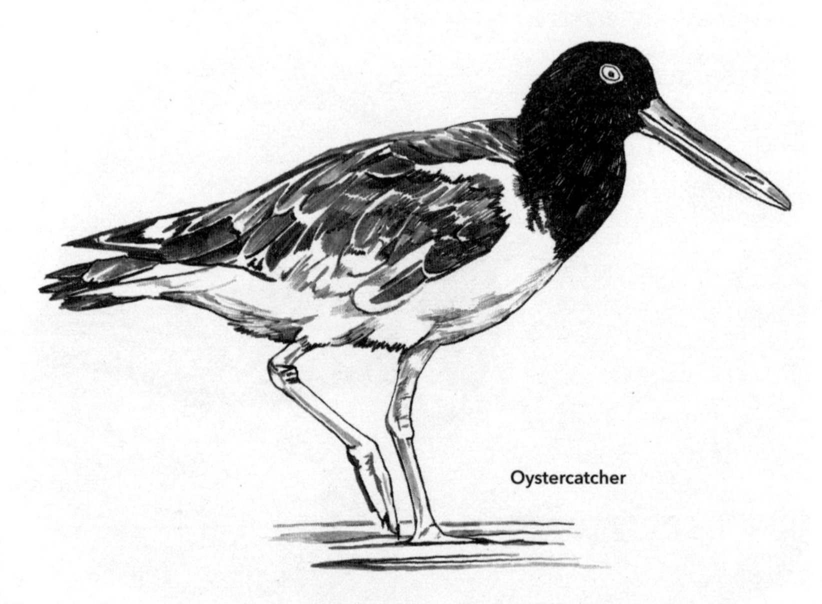

Oystercatcher

STARS OF THE SCREEN

Oystercatchers have featured a handful of times on *Springwatch*. In 2022, we watched as a pair of eggs at Wild Ken Hill in Norfolk were being dutifully incubated by both adults. It was with great excitement that we began to see the first chick hatch, taking hours to finally emerge from the egg in the early morning. However, its presence hadn't gone unnoticed and a vigilant Little Owl swooped in to make a meal of the newly hatched chick! Thankfully, the second egg hatched with more success, and the precocial chick (meaning they are able to walk from hatching) was led away from the nest by its parents.

One of the parent birds was fitted with a ring on its leg. After a bit of detective work from our wildlife camera watchers, we were able to read the unique number etched onto the ring and discover that the bird was at least 18 years old. Many wading birds can live to a ripe old age, with the oldest known Oystercatcher clocking in at 41 years, 1 month and 5 days!

WHERE TO SEE: *Oystercatchers are most easily found on the coast, where big numbers can build up in winter flocks. During the breeding season, they can also be found breeding inland, often favouring old gravel pits as spots to nest.*

WHEN TO SEE: *Oystercatchers are easily found all year round if you know where to look, but in winter their population trebles as birds arrive here from elsewhere in Europe. These large coastal flocks become a monochrome blur of raucous piping calls when they take off en masse.*

Woodcocks

The Woodcock is the strangest of all our wading birds. It has cut ties with the water and spends its days skulking through woodlands and fields, just as long as the ground is soft enough for it to probe its giant beak in to extract the worms it feeds on.

Another unusual quirk of theirs is their nocturnal lifestyle and a huge, dark bulging eye sits almost awkwardly atop their head to help them see in the dark. Their camouflage is magnificent, a mottled mix of browns, buffs and creams that give the impression it is wearing a cloak of fallen leaves. It is this that keeps them hidden when roosting during the day and safe from the gaze of predators. If disturbed, they rise from the forest floor silently into the air, not with the great clattering of a Woodpigeon or Pheasant, and are able to whisk themselves away from danger. They are built for stealth and so, although they are not a rare bird, few people have ever actually seen them.

SHINING SECRETS

Despite their incredible camouflage, a recent study led by Jamie Dunning has shown that the Woodcock possesses the brightest feather known to exist. On the underside of their tail beams the whitest white ever measured, in terms of the amount of light they reflect. At almost all times these feathers are kept hidden, so as not to ruin that immaculate cryptic plumage, and it's not quite known what exactly they use these incredibly bright feathers for. It's likely to be for some form of signalling as, just as their camouflage makes them difficult to see for predators, it also makes it tricky to spot a potential mate on the dimly lit forest floor. A few flicks of their brilliant white tail feather, however, and suddenly it's a lot easier to alert any potential suitors.

A SIGN OF AUTUMN

Whilst they breed throughout the UK, autumn sees a colossal influx of them when, under the cover of darkness, up to one and a half million birds arrive in the British countryside to escape the freezing conditions of northern Europe and Russia. Folklore tells us that the full moon closest to late October/early November is the 'Woodcock moon' when the main bulk of migratory birds will arrive.

Country folk of the past have long looked to the Woodcock's autumn invasion as a prophetic moment in nature's calendar. If the autumn has been so wet that hay hasn't been harvested by the time the Woodcocks arrive, then it is believed to spell bad news. This is compounded even more if the spring was so cold that you weren't able to sow your oats before the first Cuckoo was heard:

Cuckoo oats and Woodcock hay,
Make a farmer run away.

WHERE TO SEE: *The best place to see Woodcocks roding is in open woodland or woodland edges. Position yourself to have as much view of the sky as possible and keep your eyes trained for their fast-moving characteristic shape bombing across the fading light.*

WHEN TO SEE: *The only time Woodcocks can be reliably seen is in the spring, when males conduct display flights over their territory in a behaviour known as 'roding'. They will fly in wide arcs above the height of the trees, letting out characteristic grunts and whistles as they go. This begins at sunset, providing the chance to see them in silhouette as they do the rounds.*

Great skua

Great Skuas

On the northern tips of the British Isles, along windswept coastlines of rugged islands set steadfast against the waves, lives a notorious pirate. With a barrel chest and wingspan approaching a metre and a half, its brown body gives way to black-and-white wingtips cut from the same colour scheme as the Jolly Roger itself.

RULER OF THE SEAS

The Great Skua has earned its reputation for piracy thanks to its love of stealing a free meal from the other birds it lives alongside. Upon seeing another seabird carrying fish, the Skuas enter attack mode, harassing their target until it drops its catch. They're able to take on birds up to the size of Gannets, using the brute-force technique of grabbing them by the wing in mid-air to make them crash into the sea, where they can be attacked until they surrender their catch.

ON THE MENU

Their brutish behaviour doesn't stop at piracy, as these large, powerful birds will also readily make a meal out of other birds, particularly smaller species like Puffins as they return to their colonies. On St. Kilda in the Outer Hebrides, they've been observed conducting night raids on Storm Petrel colonies to help themselves to these little bite-sized seabird snacks.

In some areas, Great Skuas can nest in colonies of thousands. In these situations, it seems that the majority feed on fish, including those thrown away by fishing boats. It is those birds nesting at low density in small colonies that seem to favour hunting seabirds.

WHAT'S IN A NAME?

The word 'Skua' comes from the Faroese name for the Great Skua: *skúgvur*. In North America, Skuas are known as Jaegers, from the German Jäger, meaning 'hunter'. In the UK, many refer to Great Skuas as Bonxies. This word originated in Shetland and is of Norse origin and may have been derived from a description of the birds' stocky, dumpy appearance.

HEADS UP

Skuas have little fear of other predators when it comes to defending their nests. Those looking after or monitoring the areas where Skuas nest are regularly dive-bombed as the birds attempt to drive any perceived threat away!

MIXED FORTUNES

The population and range of Great Skuas have been increasing in Britain and Ireland for some time, with the birds on Orkney and Shetland believed to make up 60 per cent of the world population. Thanks to ringing, birds from Shetland have been recorded emigrating to form colonies as far away as northern Russia. However, they were hit hard by the avian influenza outbreak in the summers of

2021 and 2022 and it's yet to be revealed just what this has done to their overall numbers.

WHERE TO SEE: *Their strongholds are in Shetland and Orkney, but they have spread their breeding range south into the islands of west Scotland. Recent years have also seen them breed on Rathlin Island in Northern Ireland.*

WHEN TO SEE: *Although they can be seen around the coast all year round, many head to winter off the coast of West Africa. Spring and summer at their northern breeding colonies are the best times to see them.*

Gulls

It's a favourite birder's complaint to say, 'There's no such thing as a seagull,' when they're brought up in conversation. Whilst perhaps a bit of a nitpick, as seagull is now a well-used colloquial alternative, it is true that the Gull family contains over 50 different species across the world, some of which do not need to live by the sea at all!

Herring Gulls

This is THE seagull – the chip-stealing, trawler-following soundtrack of the seaside that has become one of the most iconic and recognisable birds in Britain. As their name suggests, Herring Gulls were once exclusively birds of coastal areas, feeding on sea fish and crustaceans whilst nesting on rocky cliffs. It is a very adaptable species and now can be found far inland, making use of rubbish dumps or the food available in towns or cities.

In adult plumage they are smart-looking birds, with a clean white head and body and light grey wings like frosted glass – giving them old names like Silver Back and Silvery Gull. Their beak is yellow with a red tip, used by the chicks as a target to peck to stimulate the adult to release food.

Despite seeming very common, the Herring Gull is actually a red-listed bird due to it suffering recent steep declines in population. Whilst urban populations seem to be on the rise, this hasn't made up for dramatic decreases in the coastal breeding populations, which has meant there are now estimated to be half as many Herring Gulls.

WHERE TO SEE: *Large urban colonies are present in cities such as Cardiff, Bristol, London and Aberdeen, whilst almost all of the British coastline offers the chance to see this iconic bird.*

WHEN TO SEE: *All year round.*

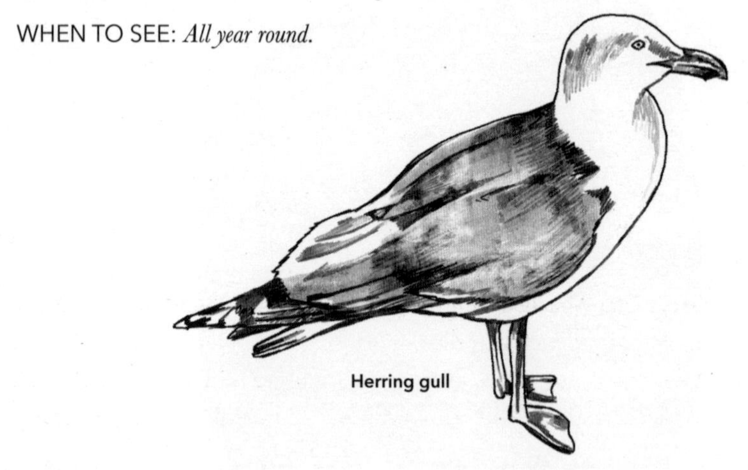

Herring gull

Great black-backed gull

Great Black-backed Gulls

The largest gull in the world, with a wingspan of over 1.5 metres, the Great Black-back is a real titan of our seaside skies. Described as 'the King of the Atlantic', their size and strength make them efficient scavengers, aggressive hunters and fearsome pirates.

In parts of Ireland, it was given the name Goose Gull because of its size, and it is actually larger than some of the goose species that visit the UK, like Barnacle and Brent Geese. Its scientific name *Larus marinus* translates to 'sea gull' and, given it is restricted almost entirely to coastal areas to breed in the UK, it is perhaps the most deserving of the famous name.

APEX PREDATOR

Great Black-backed Gulls regularly take fish found at the surface of the sea as well as crustaceans, sea urchins, starfish and squid. They're also not above joining other gulls to feed at rubbish dumps, particularly in the winter months, but what really sets them apart is their highly predatory behaviour. They're an apex predator within seabird colonies and feast on eggs, chicks and fledglings of all other seabird species around them, even other gulls. It's not only young birds that they will target, however, and Great Black-backed Gulls will regularly hunt smaller adult seabirds like Puffins, Shearwaters and Terns. On Skomer Island off the Welsh coast, rabbits make up an important part of their diet and they have even been seen to swallow them whole.

WHERE TO SEE: *Great Black-backed Gulls are best looked for in coastal areas, particularly islands with large breeding populations of seabirds like Skomer and Skokholm. The Northern Isles, northwest Scotland and the Atlantic coast of southwest Ireland host the highest breeding densities.*

WHEN TO SEE: *All year round.*

Black-headed Gulls and Mediterranean Gulls

If there is a gull that you may regularly encounter without living anywhere near the coast, then it is the Black-headed Gull. They nest in large, noisy colonies that can be found on islands or spits of land on our most inland water bodies. In winter, up to 2 million of them are estimated to arrive from the Continent, when they will roam our towns and countryside in search of food.

The inland habits of Black-headed Gulls have resulted in names such as Pickmire, Picktarn and Carr Swallow, which reflect the inland waterscapes that they can be found in. They are much smaller than the gulls mentioned previously and have been given names like Pigeon Gull or White Crow that reflect their size but potentially also their fondness for living alongside people.

NEW NEIGHBOURS

Although called the Black-headed Gull, their heads in the breeding season are really a warm chocolate brown colour. There is another gull more deserving of the title that began breeding here in 1968 and has since expanded its range rapidly. The Mediterranean Gull now breeds in all four countries of the UK, regularly joining breeding colonies of Black-headed Gulls. The Med Gull (as they are known by birders) has a jet-black head in the summer months, with a bright, blood-red beak and legs. In the winter, both species lose their hoods and have largely white heads to see them through to spring.

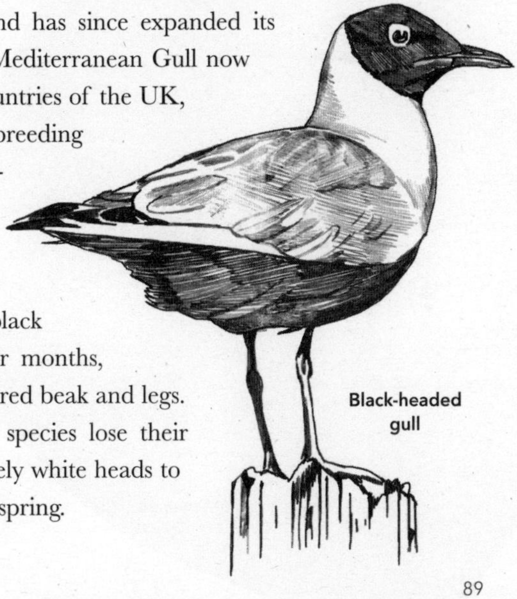

Black-headed gull

WHERE TO SEE: *Everywhere! From car parks to coastlines, the Black-headed Gull can turn up anywhere, especially in winter. Belmont Reservoir in Lancashire is probably the biggest colony in the UK. Mediterranean Gulls are often found within Black-headed Gull colonies, particularly in the south and southeast at places like Langstone Harbour in Hampshire.*

WHEN TO SEE: *In the breeding season, both species are tied to their breeding colonies but will wander much wider in winter. This is also when it's more likely to see Mediterranean Gulls ranging further north in greater numbers.*

Kittiwakes

Anyone visiting a seabird colony can be in no doubt about how the Kittiwake got its name. The sound of their 'kitt-i-waake, kitt-i-waake' call is deafening when faced with a chorus of thousands, their cries reverberating off the steep cliffs they build their precarious nests on.

Kittiwakes are only seen in the spring and summer when they come to land to breed. The rest of the year they spend wandering the high seas of the Atlantic Ocean, where they forage at the water's surface for small fish.

Kittiwakes have a particular taste for sandeels and over-harvesting of these fish by humans can lead to dramatic effects on Kittiwake numbers. Sandeels are also particularly sensitive to changes in ocean temperature, preferring cooler waters, and there's a real worry that warming oceans will cause their populations to move further north and impact on the health of Kittiwake populations.

CITY LIVING

Colonies are spread around the UK coastline, but the highest densities are found along the East Coast, particularly between Flamborough Head and Orkney. Usually, sea cliffs are the chosen nest sites, although manmade structures are increasingly being used. The River Tyne is home to the most inland breeding colony of Kittiwakes in the world, at about 12 miles up river from the sea. The buildings and bridges of Newcastle and Gateshead have proved the perfect substitute for the Kittiwake's natural sea cliff homes, and their urban population has continued to expand at a time when many other colonies have decreased.

WHERE TO SEE: *Kittiwakes are often associated with other seabird colonies. RSPB Bempton Cliffs on the Yorkshire Coast possesses one of the largest mainland colonies; whilst Orkney holds almost 20 per cent of the UK population. Rathlin Island in Northern Ireland and Skomer in Wales have the largest colonies in their respective countries.*

WHEN TO SEE: *Kittiwakes return to their breeding colonies in spring. Aim for May or June to get the full sights and sounds of the colony!*

Arctic Terns

The Arctic Tern is the Olympic champion of the animal world. When it comes to migration, no other species on Earth travels as far each year to get between their breeding sites and wintering spots.

RECORD BREAKERS

In 2016, an Arctic Tern on the Farne Islands off the coast of Northumberland claimed the title of the longest migration ever recorded when a tracker recorded it covering almost 60,000 miles to Antarctica and back. The 100g bird left the Farnes in July and one month later had made it to the tip of South Africa. It spent most of October in the Indian Ocean before arriving at Antarctica in February, where it stayed until it began its northward migration in the spring.

The meandering migration route taken by Arctic Terns is equivalent to twice round the circumference of the Earth, and it's believed in their long lifetime of up to 30 years they will travel the distance of almost four trips to the moon and back. What's even more extraordinary is that they will undertake these mammoth journeys whilst only very young. In 1982, a chick ringed on the Farne Islands in the summer was found in Melbourne, Australia, only 3 months after it had fledged the nest.

CHASING THE SUMMER

Because of their impressive migration, Arctic Terns are said to be the animal that sees more daylight than any other as they journey from pole to pole. They mainly eat small fish, which they catch through diving into the water and swooping low to pick them from the surface. Finding their preferred small prey in a Northern Hemisphere winter is hard. The ocean surface tends to be calmer in the summer, and a

Arctic tern

sea that is being tossed around by strong gales makes it difficult to find the small creatures that the terns feed on.

They're able to travel such epic distances thanks to their light build and long wings, meaning they can glide effortlessly with very little flapping. Their forked tail allows them to be extra agile when hunting and has given them the alternative name Sea Swallow. Most Arctic Terns spend the non-breeding season out to sea and will sleep and eat whilst on the wing as they are unable to swim well and so avoid settling on the ocean surface like other seabirds.

UNDER ATTACK

The Arctic Tern is by far the commonest breeding tern species in Britain and Ireland, but their northerly distribution and exclusively coastal nesting habits mean they're not a bird that many people will be overly familiar with. Visiting a tern colony is a raucous experience, with their loud piercing calls filling the air. If you find yourself at one of the colonies where you're able to walk on designated footpaths between the tern nests, then a hat is a must!

Arctic Terns are one of the most aggressive species when it comes to protecting their nests and will attempt to drive away any perceived intruders. Prepare to be ruthlessly dive-bombed if setting foot on any island they call home, with spikey pecks courtesy of their sharp beak and an ear full of rattling calls being a regular occurrence. In breeding colonies much further north than the UK, they've even been seen driving polar bears away from their nest sites!

WHERE TO SEE: *The most infamous population of Arctic Terns is on the Farne Islands, where a walk through the colony will see you receive a fair few pecks to the head. In recent years, the island has been closed to visitors due to the risk of spreading bird flu, but colonies can still be seen at Cemlyn Bay in Anglesey, Copeland Island in Northern Ireland and Forvie National Nature Reserve in Scotland.*

WHEN TO SEE: *Arctic Terns are summer visitors, so catch them over spring and summer before they head to the Southern Hemisphere to spend the winter.*

Puffins

Puffin

You don't have to like birds to love Puffins. They are one of the most unanimously loved animals in the world, even though most people have never seen one. It's not hard to see why they have such an adoring fanbase. Their brightly coloured beaks and feet stand out against their black-and-white bodies, giving the impression of a little penguin that decided to join the circus. The way they waddle is cute, the way they frantically flap their tiny little wings to fly makes us go 'awwhhh' and the way they affectionately touch their beaks together is very wholesome too. Oh, and they make silly groaning noises, they pack their beaks with a ridiculous number of fish AND their round, fluffy chicks are called 'pufflings'. It's no wonder they're so popular.

ON THE MENU

Puffins used to be valued for a slightly different reason not so long ago, as they were a source of food for many island and coastal communities. They were easy to catch and incredibly plentiful, meaning they could be dried and salted to eat throughout the lean winter months. They were one of the birds that, like the Barnacle Goose and some other seabirds, were believed to be fish rather than birds, meaning they were exempt from the ban on eating meat during Lent.

STACKING SANDEELS

Puffins nest in burrows, so need areas where they can dig into the ground to nest. To make the most of their fishing trips out to sea, they try to carry as many sandeels in their bill as possible when returning home. The two mandibles have backwards-facing serrations on them and are hinged in a way that means they can be holding fish in their beak whilst still actively hunting more. The record for the highest number of sandeels ever counted seems to be disputed – with some sources stating 61 and others claiming 83!

SECRET SIGNALS

Puffins' beaks are their most eye-catching feature but may not look the same to Puffins as they do to us. When a UV light is shone on them, we can see that their beaks fluorescently glow. Puffins don't need the UV torch, however, and are able to see these characteristics during their day-to-day lives. But what exactly is the reason behind this secret signalling?

Well, the bill of a Puffin doesn't look the same all year round. At the end of the breeding season, parts of the bill actually drop off, leaving a less broad, duller and overall less extravagant-looking beak. We rarely see Puffins looking this way given that they spend their winters out on the high seas, only returning when it is time to breed – with their fancy summertime bills already developed. The fact that their bills grow in time for the breeding season shows they must play an important role in attracting a mate. Our best assumption is that the UV details of a Puffin's beak help them look extra sexy in their search for a mate.

WHERE TO SEE: *Sites ranging from the Farne Islands, Bempton Cliffs, Rathlin Island, Skomer and Sumburgh Head to the Isles of Scilly show how Puffins can be found at various sites all around the UK coast. The island of Lundy in North Devon preserves the Norse word for Puffin: lunde.*

WHEN TO SEE: *April to July is when Puffins are best seen as, once breeding is over, they will return out into the Atlantic Ocean to spend the winter bobbing about on the open seas.*

Pigeons and Doves

Pigeons are one of the few birds that you can reliably bet on encountering every single day. Whether in the countryside or the city, pigeons are everywhere. But their ubiquity should not mean we overlook them, for they are some of the most interesting birds around.

Before we meet our species, however, let us address something that's a little misleading – the words 'pigeon' and 'dove' themselves. There really is no need for both of them, as there is no biological difference between pigeons and doves at all – they are the same group of birds. We tend to call smaller, daintier pigeons 'doves' but many other languages don't differentiate between the two. It seems whether you were labelled a pigeon or a dove just came down to whether you had a good PR team!

Feral pigeon

Feral Pigeons

There's probably no more recognisable bird in Britain than the Feral Pigeon. A visit to any major town or city wouldn't be complete without their flocks, lining the rooftops and descending to the pavements for scraps, being scattered by busy pedestrians on their way to work.

BAD REPUTATION

Few animals are held with such contempt globally as the Feral Pigeon. They're a universally unpopular bird, a pest, a nuisance – something to be fought with spikes to keep these 'rats with wings' away from the places we want to be. They've been demonised as a dirty, disease-spreading, destructive enemy to be despised at all costs – the manifestation of the wild coming into our human spaces and making them 'unclean'. But the humble Feral Pigeon is a bird with an incredible story, and one that we humans owe a lot to.

A HUMAN HISTORY

They are the oldest domesticated bird in the world, with Mesopotamian tablets mentioning the domestication of pigeons over 5,000 years ago. Some sources even suggest we may have been breeding pigeons thousands of years earlier still.

As a source of food, they were easily kept and required minimal care. Dovecotes have been built throughout the ages across the world to provide them with safe places to nest so that eggs and young birds could be harvested for food. Droppings could be utilised too, spread on crops for fertilisation or used for leather powder and gunpowder making. The oldest dovecotes are believed to be from the Middle East and North Africa, with the oldest evidence in Britain found in the 'pigeonholes' cut into the Roman settlement of Caerwent in Wales.

ANCIENT ANCESTORS

So, if the Feral Pigeon is a domesticated bird, then who is its wild ancestor?

Enter, the Rock Dove. This wild ancestor of the Feral pigeon lived on cliffs and rock formations around Europe, North Africa and parts of South Asia. The ancestral Rock Dove doesn't come in the same myriad of colours as its city cousin, preferring to stick to the grey back and black wing bars that are still a common sight on the streets.

It was the Rock Dove that was taken into these ancient dovecotes. Through the domestication process, they were bred in a whole mix of greys, blacks, browns and whites. In fact, pure white 'doves', the symbols of peace and the ones frequently released at weddings, are just a colour morph of these birds.

Their use meant they were moved all around the world and it wasn't long before they found their way into the wild. Buildings in towns and cities replicated the cliff faces that their ancestors had favoured in the past and so the Feral Pigeon was born.

Collared Doves

Wind the clocks back to the 1950s and the Collared Dove wasn't anywhere to be seen on British shores. In fact, in the early 1900s, you could barely even find them in Europe – with Turkey being their only outpost on the continent.

EURO TRIP

They have their origins in India, where they'd lived happily for thousands of years before they began to show signs of spreading. What nobody could have predicted, however, was their sudden continental-wide conquering of Europe. They arrived in Hungary in 1932, Germany in 1945, the Netherlands in 1947 and France and Switzerland by 1952 before reaching British shores in 1955. Their spread was astonishing, calculated at about 27 miles per year, and by 2000 they could be found everywhere from the Arctic Circle of Norway to the Canary Islands off the coast of North Africa.

MAKING THEMSELVES AT HOME

Exactly what triggered the Collared Dove to spread so enthusiastically is still unknown. We know they're a species that's happy to live in our human-modified landscapes, and their unfussy diets mean there's always plenty of seeds and plant material for them to eat. Like most members of the pigeon family, they also breed rapidly and they are one of the few British birds recorded nesting in every month of the year. All these things, coupled with their strong flight and ability to disperse over a large area, have resulted in the Collared Dove becoming a familiar sight and sound to most of us in the British Isles.

WHERE TO SEE: *They seem to prefer towns and villages and avoid intensely urban areas. Large numbers can gather in gardens or farmyards where food is on offer.*

WHEN TO SEE: *Any time of year.*

Turtle dove

Turtle Doves

Turtles Doves are smaller, not much larger than a Blackbird. With a chest of warm pink, a bright red eye and legs and orange-and-black scalloped feathers that sit across their wings like ornate chainmail, they possess an exotic aura. Unsurprisingly then, the Turtle Dove is the only completely migratory species of pigeon found in the UK, disappearing from our shores in late summer to spend the winter in Africa.

WELCOME RETURN

As with other African migrant birds, their return is seen as a symbol of spring, their soft, purring, rhythmic call a true tonic for the soul and an idyllic soundtrack to lazy summer days. The Song of Solomon in the Old Testament of the Bible describes this herald of spring:

> *The winter is past, the rain is over and gone. The flowers appear on the earth; the time of singing has come, and the voice of the turtle dove is heard in our land.*

WHAT'S IN A NAME?

It's this call that gives the Turtle Dove its name. Not because it sounds anything like the marine reptile, but because the call itself sounds like a repeated, cooing 'tur-tur' sound. Over time, the birds became known as turtles/tortles/turtels/turtuls, until we reached the name that we have today.

LOVE BIRDS

When it comes to our perception of Turtle Doves though, perhaps the thing we associate with them the most is romance and love. Turtle Doves are almost always portrayed in pairs – most famously in the Christmas, carol 'The Twelve Days of Christmas' where two Turtle Doves are given to the singer by their true love.

Shakespeare used them regularly in his works too, giving them mentions in some of his famous plays when he needed something to symbolise devotion. Turtle Doves are still used to represent romance today, with lines in songs by Cliff Richard, Annie Lennox, Barry Manilow, David Gray, Elvis Presley, Frank Sinatra, Journey and Madonna all including references to the bird.

LOVED AND LOST ...?

The sad reality is that the Turtle Dove is fading from our land. Since 1970, there's been a 98 per cent decline in their numbers, making them the UK's fastest-declining species. Loss of habitat, increased use of herbicides that kill off the weeds whose seeds they rely on and unsustainable levels of hunting along their migration route have put them on the brink. Conservation projects are working hard to address the issues they face, and only time will tell whether the purring of the Turtle is consigned to stories and songs.

WHERE TO SEE: *Their remaining strongholds are largely southern and eastern England, at sites such as the Knepp Estate in West Sussex and Snettisham Coastal Park in Norfolk. The Sutton Bank National Park Visitor Centre in the North York Moors is one of the best spots to see them in northern England.*

WHEN TO SEE: *These summer migrants arrive from April to croon their soft songs. By September, most have left our shores to journey south to Africa.*

Stock Doves

Stock Doves are tidy little birds. With a neat blue-grey sheen and glossy colour, they look very similar to a Woodpigeon, although one that's been through the wash and shrunk a few sizes. They like big old trees or barns to nest in, so can often be found in farmland, where they also have access to readily available food. Their main feature to distinguish them from their bigger brother is the lack of white on their collar or wings.

Stock Doves are more readily encountered by sound rather than sight. They project a deep, disyllabic 'woo' sound in the spring, mistaken by some for the hooting of an owl.

WHERE TO SEE: *Stock Doves are generally most common on farmland, where they can often be seen feeding in stubble fields alongside Woodpigeons. They will also visit rural gardens to feed.*

WHEN TO SEE: *They're most easy to find in spring, when their powerful song makes them hard to miss. It's sometimes mistaken for the hooting of an owl but follow the sound to a hollow tree or barn and you'll see its pigeon perpetrator!*

Woodpigeons

Hulking round our gardens and parks like a bouncer, the Woodpigeon has become even more familiar over recent years as it has increasingly got to grips with urban life. There are over 5 million pairs estimated to be living in the UK, having undergone a population explosion of around 150 per cent over the last 50 years.

Their success, like the Collared Dove, lies in their adaptability and ability to breed all year round. Looking at the nest of a Woodpigeon, you'd be hard-pressed to think that this is a bird that could ever successfully raise a chick. If you manage to find yourself beneath a Woodpigeon nest, a mass of thin twigs thrown haphazardly on top of each other in a bush like a toddler playing Jenga, then make sure to look up. The sticks are stacked to make a platform so thin that you'll be able to see daylight through it. How the rather hefty weight of an adult Woodpigeon, let alone two well-grown chicks, can be supported by it is a mystery, but the evidence for their success is all around us!

WHERE TO SEE: *Almost anywhere. There are not many places in Britain where you'll struggle to find a Woodpigeon!*

WHEN TO SEE: *Although they can be seen all year round, look for large flocks that gather over the winter. Large numbers migrate to our shores from the Continent in autumn, sometimes in flocks thousands strong!*

Woodpigeon

Cuckoos

There aren't many birds that can stake a claim to being as recognisable by their song as the Cuckoo. Immortalised in clocks since the early 1700s, the male's song is a sign that spring is in full swing.

CUCKOO CUSTOMS

Hearing the first Cuckoo of the year has held great significance throughout the ages, and a whole selection of practices has been built up around it. Turn the money in your pocket and spit on it if you wish it to stay plentiful throughout the year. Some people would roll in the grass to protect themselves from back problems, others would pay attention to the ground beneath their feet. A hard surface meant bad luck in the coming year, but soft and grassy indicated all would be well. In the Hebrides, it was deemed bad luck to hear a Cuckoo whilst you were hungry and was a surefire sign of a bad harvest to come.

Girls would check their left shoe upon hearing the Cuckoo's song, expecting to find a hair the same colour as that of their future husband, whilst Pliny the Elder wrote that the soil immediately behind your right foot at the time of hearing the Cuckoo could be gathered and used as a flea repellent.

DID YOU KNOW?

Before migration was well understood, some didn't believe that the Cuckoo left these shores at all, but instead morphed its form into a hawk. It might seem an outlandish idea these days, but it's based on a striking similarity between the body shape of Cuckoos and hawks. Like everything in nature, this is unlikely to be coincidental. Studies have shown that mimicking the appearance of hawks and appearing as a threat to the smaller birds they parasitise the nests of means they're more likely to scare them away from their nests to lay their eggs unwatched.

As Cuckoos repeat their phrases over and over again in their song, another tradition built up around the number of times they would call before falling silent. People would ask, 'Good bird, how many years before I die?' and the Cuckoo's chimes would give you the answer. Another variation would involve shaking a cherry tree at the same time to be told how many years it would be before you were married.

A BIRD LIKE NO OTHER

Many are likely to be more familiar with the sound of a Cuckoo than its appearance, but it is a striking bird to behold. They're around the size of a dove with a long, tapered shape that gives them a predatory air if you see one in flight. Their colours are subdued but smart, dressed in a blue-grey shade on their neck, wings and head. This is in contrast to their starkly barred underparts, which are so striped they almost resemble a comedy convict. Then there's the piercing yellow eye, staring out of the greys and whites with an aura of the slightly unhinged.

DEVIOUS DECEIT

Cuckoos are infamous for the way in which they go about raising chicks – in that they don't play any part in raising them at all! Female Cuckoos seek out other birds' nests to lay their eggs in. A single female can lay up to 25 eggs in a single spring, all in different nests. Female Cuckoos generally specialise in parasitising one species and have evolved over time to be able to replicate the colour and pattern of the host's eggs so that their devious plan isn't rumbled.

Cuckoos that have been parasitising a particular species for longer have eggs that match them more closely. For example, Cuckoos do a pretty good copy of a Reed Warbler egg as, over the generations, their hosts have got better at spotting the fakes. Dunnocks, on the other hand, are a much more recent target for Cuckoos and so haven't quite clocked on yet that the egg that looks nothing like theirs is a problem. They will do soon though, and the evolutionary pressure for the Cuckoo to begin adapting a matching egg will begin, kicking a new evolutionary arms race into gear.

Cuckoo

When the Cuckoo chick hatches, its very first instinct is to push all other eggs or fellow chicks out of the nest, securing all the parents' attention entirely for itself. Driven by the instinct to care for the begging chick they find in the nest, the foster parents continue to ply it with food, seemingly none the wiser that their efforts are being put into an imposter.

FLEETING VISITS

We like to think of Cuckoos as British birds, but adults often only spend around 6 weeks on our shores each year. Given their devious egg-laying strategy, they are able to start their journey back to warmer climes as early as mid-June and are long gone by the time July ends.

An old rhyme to remember their cycle goes:

The Cuckoo arrives in April,
He songs his song in May,
In the middle of June, he changes his tune,
In July he flies away.

BASIC INSTINCT

The Cuckoo's chicks are left behind, being doted on by their foster parents who shovel copious amounts of invertebrates into their huge gape until they're ready for their own independence. The fact that these young birds, having never met another Cuckoo in their lives, know when, where and how to migrate is one of the most astonishing examples of instinctual behaviour in birds.

GOING CUCKOO

The Cuckoo, with its fascinating behaviours and mastery of the dark arts, has stitched itself into the fabric of folklore wherever it is found. One of the most curious examples is how the bird's name became a descriptor of those deemed to have taken leave of their senses. Greek playwright Aristophanes was the first to invent Cloud Cuckoo Land in his 414 BC play *The Birds*. In the story, a man from Athens persuades the world's birds to create a new idyllic city in the sky. Since then, the phrase has gone on to be used to describe something unrealistically idealistic. It may have been from here that we got the word 'cuckoo' to describe someone perceived as crazy, or that may be from the enthusiastic way that the bird repeats the same notes over and over again for hours on end!

WHERE TO SEE: *Cuckoos can be found across a wide variety of habitats but can most easily be seen on large wetland reserves, open woodland or moorland with scattered trees. Try areas such as Bannau Brycheiniog in Wales, West Highlands of Scotland, Dartmoor in England and Ballynahone Bog in Northern Ireland*

WHEN TO SEE: *The best time to spot a Cuckoo is when they're making themselves easy to find. Head out at daybreak in April or May to enjoy a springtime dawn chorus and listen out for that famous, far-carrying sound.*

Owls

Everybody loves owls. Immortalised in storybooks and folklore for centuries, there's no doubt that the owls are firm favourites of ours.

Their popularity is in direct contradiction to how often we actually *see* owls. When was the last time you laid eyes on one? The iconic hooting of the Tawny Owl is likely to be the most familiar to us or perhaps you may have been lucky enough to catch the ghostly white apparition of a Barn Owl in the periphery of a car's headlights. They're not a bird that reveals themselves to us often, making each encounter all the more special.

WISE OR LIES?

One of the most recognisable avian phrases is 'the wise old owl'. Their wisdom has been hailed at least as far back as Ancient Greece, where Athena, the goddess of wisdom, had an owl symbol.

It's easy to see why they get this moniker, sitting silently stoic all day, as though deep in philosophical thought. Their round faces, forward-facing eyes and tiny, nose-like beak make them the most human-looking of any bird. Unfortunately, the intelligence of owls in a classic ornithological myth.

Owls don't have particularly large, well-developed brains and haven't shown any great levels of intelligence when tested. In fact, most of the space in their skull is devoted to their huge eyes, which give them their extraordinary night vision capabilities.

SILENT BUT DEADLY

Another superpower shared by owls is their silent flight. Comb-like serrations on the leading edge of their flight feathers break up the flow of air that would usually create the sound of a wingbeat and the soft, velvety texture of their feathers further dampens any sound.

THE FAB FIVE

In Britain, we're blessed with five species that can be found regularly breeding in these isles, each one of them supremely adapted for the niches they inhabit. As well as the Barn, Little and Tawny mentioned below, there are the Short-eared and Long-eared Owls that tend to dwell on moors, grasslands and overlooked plots of woodland. Despite their names, the tufts of feathers that poke up from their heads aren't ears at all. The ears of owls, like all birds, are hidden openings on the side of the head buried beneath the feathers. The disc-shaped face of owls is designed to capture as much sound as possible and funnel it to those ears.

Tawny Owls

Tawny Owls are the most widespread species in the UK, although they are absent from Ireland. They're our largest owl, with a wing-span that can reach over a metre wide, but that doesn't make them any easier to see. Although they frequent most woodlands and parks, their mottled brown feathers make for excellent camouflage when they're roosting during the day.

A DUET IN THE DARK

Whilst other British species make a whole variety of noises, from shrieks to yips to low, deep booming, it's only the Tawny responsible for that iconic 'twit twoo' sound.

The call itself isn't made by a single bird but is a duet between a male and a female. The hooting 'twoo' part of the call is the territorial part made by the male, whilst the female responds with a sharp 'kee-wick' sound.

Tawny owl

DID YOU KNOW?
Other birds can be a great help when it comes to finding Tawny Owls. Listen for raucous alarm calls coming from the trees as smaller birds will often mob roosting Tawny Owls in a bid to drive the predator away.

AUTUMNAL AGGRESSION

A good time to listen for Tawny Owls is in the autumn. This is when young birds are being kicked out of their parents' territory and are dispersing from the area in which they were hatched to find a territory of their own. Adults are fiercely territorial and don't take lightly to intruders on their patch, so it's not uncommon to hear them shouting at each other during autumnal nights!

WHERE TO SEE: *Other than Ireland, Tawny Owls are found over much of the British Isles but seeing them is a different matter! Looking for them in parks and gardens where there may be some artificial light is often the best chance to catch a glimpse.*

WHEN TO SEE: *They are almost exclusively nocturnal and only active in the daytime if disturbed, so heading out at night is the best chance to track them down.*

Little Owls

The most recent addition to Britain's owl collection is the Little Owl. Although widespread in Continental Europe, it was officially introduced in the late 1800s, when 40 were set free in Kent, followed by another release in Northamptonshire a few years later. Around 60 years later, Little Owls had established themselves across much of England and Wales. Given that they only came from across the Channel, where they lived alongside most of our native species already, the Little Owl seemingly slotted in without too much drama.

Little owl

SMALL APPETITE

They like open land with trees or buildings for them to nest in. Like other owls, they will regularly hunt small mammals but it's not the main part of their diet. Unsurprisingly given their size, they spend a lot of time hunting insects and other invertebrates, particularly beetles and earthworms. If you're ever lucky enough to find a pellet from a Little Owl, you'll often notice them studded with shiny pieces of beetle like some macabre Fabergé egg!

LITTLE BIRD, LOFTY STATUS

Given the Little Owl's European origins, much of their historical folklore is found on the Continent. In Ancient Greece, they were sacred to Athena, the goddess of wisdom, and even bear her name in their scientific name *Athene noctua*. Little Owls were a common sight in Athens and were depicted on the silver tetradrachm coin, which became known as an 'owl'.

STARS OF THE SCREEN

Little Owls are not a species that has made regular appearances on the *Springwatch* cameras, but the most memorable family was one we followed back in Sherborne in 2018. Here we had a pair of owls raising their chicks in the barn right by our studio. A pair of Blackbirds had also chosen to make a nest in the same barn, a dicey place when there's a predator on the hunt.

With our cameras rigged around the barn to catch the drama, we watched the daily battles that occurred at dawn and dusk when the waking hours of the two families overlapped. The Blackbirds would relentlessly mob the owls, dive-bombing them in a constant attempt to drive them out of the barn and away from their chicks. The Blackbird chicks, growing larger in their nest, were the subject of much speculation about their fate and whether they would be noticed by the owls. It wasn't until the first Blackbird chick fledged onto the floor of the barn that the Little Owls suddenly took notice and snatched one during one of the live programmes.

The Little Owls helped themselves to two of the newly fledged chicks before they were able to escape the barn. In a further twist, it then transpired that there was another Blackbird nest in a barn directly across the courtyard! These chicks were much younger, but the owls had clearly got a taste for Blackbird and were soon carrying them back to their nest to feed to their own offspring.

It may have been a premature end for the Blackbirds, but it helped our Little Owl chicks go from strength to strength. Their mother was a fearsome hunter and the chicks were never left to go hungry. As well as Blackbird chicks, their larder was always well stocked with all manner of prey items, including other birds, mice, voles and even bats!

WHERE TO SEE: *Most abundant in East Anglia, the Midlands and northern England, look for Little Owls in open farmland and parkland with old, hollow trees for them to nest in.*

WHEN TO SEE: *The best time to spot them is in spring and summer when the long days mean they can be seen in the daytime in the early morning or evening.*

Barn Owls

A ghostly pale shape floating through the midnight air can only be one thing.

Barn Owls can go toe-to-toe with most other species when it comes to vying for the title of the nation's favourite. Their striking white fronts and heart-shaped faces make them easily recognisable, and their habit of hunting during the daytime makes them easier to encounter than other species of owl.

Before humans, they naturally lived in caves and hollow trees, but Barn Owls have now lived alongside us in our farms and villages for centuries, making use of our barns and abandoned farm buildings.

HAUNTING THE NIGHT

They've got a plethora of old names, many of which allude to the superstitions surrounding them. Banshee, Ghost and Demon Owl are unsurprising considering their ghostly, wraith-like apparitions in the dark. Barn Owls have relatively long wings for their weight, which allows them to fly at an almost supernaturally slow speed – as little as 2 miles per hour. This gives the impression they are of floating over the fields, a phantom roaming the midnight air. In Gaelic-speaking Scotland, they were known as *cailleach-oidhche gheal* – 'the white woman of the night'.

To add to this ethereal aura, they also possess a piercing shriek, which cuts through the still night air and gives them names like Screech Owl, Hissing Owl and Roarer. During the eighteenth and nineteenth centuries, it was said that the sound of this screech by the window of someone who'd fallen ill was a sign of imminent death.

IMPROVED RELATIONS

It used to be believed that a dead Barn Owl nailed to the door would ward off evil spirits. Their eggs could be turned to ash to be used in a potion to improve eyesight, whilst an owl broth was given to children who suffered from whooping cough.

Thankfully, the Barn Owl enjoys a much better relationship with people these days. Although the renovation of farm buildings has resulted in some loss of nesting sites, many farmers install nest boxes on their land to encourage Barn Owls to make their home. Not only does the sight of the birds bring joy, but their hunger for rodents means they're often referred to as 'the Farmer's Friend'.

WHY SO WHITE?

It might seem a bit of a strange tactic for a bird that hunts at night to evolve to be glaring white. Especially when you consider all the effort that's gone into giving them their extraordinary powers of silent flight to allow them to sneak up on their prey undetected. To explain why Barn Owls seem to have this paradoxical colour scheme, we have a study published in 2019 (San-Jose, L. M. et al.) to thank. Although the vast majority of Barn Owls in the UK are white, many in Europe have darker, buff-brown fronts instead. The scientists looked at Barn Owls with different colourings to assess their success levels when it came to catching a meal. They found that Barn Owls with white fronts were more successful at hunting on moonlit nights, when they were at their most visible compared to their brown-coloured brethren. They explained this by demonstrating how mice and voles will freeze upon seeing the white plumage, giving the owls longer to make the kill. So, whilst it may seem counterintuitive at first, the owls are actually benefitting from making themselves more visible!

STARS OF THE SCREEN

We met one of *Springwatch*'s most beloved characters back in 2011 at RSPB Ynys-hir in west Wales. Bob was the smallest of four Barn Owl chicks and there was real worry that he was in danger of becoming a snack for his older siblings. Many birds of prey incubate their eggs so that the chicks hatch asynchronously, meaning they hatch on different days and therefore are different ages. This is a strategy to ensure that, in the event of a food shortage, the older, stronger chicks are prioritised, with the youngest chicks sometimes even becoming food themselves! In fact, in 2006, *Springwatch* featured a Barn Owl nest of

7 chicks in the Lost Gardens of Heligan in Cornwall where one of the larger ones ate its smaller sibling alive!

Thankfully there was no such end for Bob in 2011, as an abundant year for voles meant the parents kept the chicks well fed as they grew. Bob's only problems came instead from how to eat all the food!

WHERE TO SEE: *Barn Owls can be found in most farmland landscapes across Britain, with their highest densities in the flat lands of East Anglia. Look for them quartering over rough grassland and field margins as they hunt for rodents.*

WHEN TO SEE: *Barn Owls can be seen hunting in the day throughout the year. You increase your chances of a daytime sighting by being out at dawn or dusk.*

Nightjars

The Nightjar is one of the most ethereal birds to haunt the British Isles. Everything about them is unusual. Their camouflage is extraordinary – a vermiculate mix of browns, born of wood and earth. It gives them the appearance of an animate log or branch, coming to life with an eerie, huge, alien eye that widens as dusk falls and the bird begins to stir.

OPEN WIDE

When the Nightjar takes to the air at nightfall, it utilises another alien feature. They may possess a tiny beak, but they can open this to reveal a colossal mouth that wouldn't look out of place on the front of a whale. They use this to catch moths, hawking them on the wing over heathlands and woodland clearings like a giant, ghostly swallow.

SCAPEGOATS

It's their mouth that gives them their most well-known alternative name: the Goatsucker. The story had it that Nightjars would latch their mouth around the udders of goats to drink their milk, depriving their offspring, or the farmer for that matter, of any milk. In addition, it was said that any goat that fell victim to a Nightjar would stop producing milk and may even go blind.

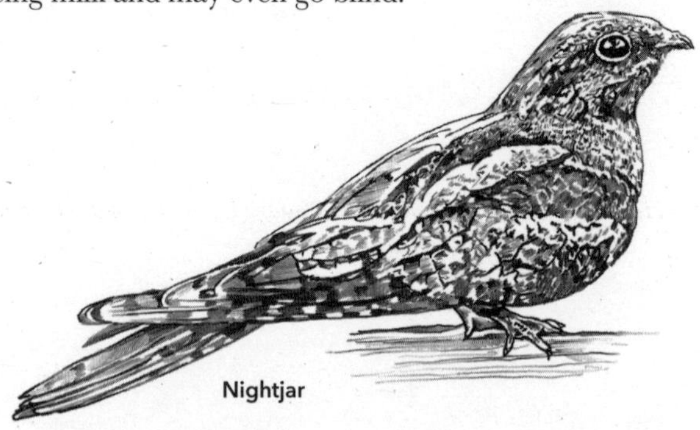

Nightjar

As long as 2,000 years ago, the Greek philosopher Aristotle was writing of the bird's well-known love of goat's milk and this belief persisted for centuries. Eighteenth-century naturalist Gilbert White wrote of his frustrations that it only took 'the least observation and attention' to realise that the Nightjar was simply eating insects. To this day, its scientific name *Caprimulgus* translates to 'Goatsucker'.

Shepherds of the past would have routinely encountered Nightjars as they grazed their animals on open ground. The presence of flying insects around the livestock would have attracted Nightjars to feed as darkness fell, with their strange silhouettes wheeling and swooping around the herder's precious animals. Anyone who found a dead Nightjar would have likely been perplexed that their tiny beaks were not much good for eating anything, but that their huge mouths were perfectly udder sized.

A SOUND LIKE NO OTHER

If it wasn't enough that the Nightjar looks and behaves unlike any other birds of these isles, its song also has no equivalent. Warm summer evenings are filled with the sound of rhythmic churring that continuously bubbles and rattles for minutes on end. Its old names – Night Churr, Eve Churr and Churn Owl – speak to its otherworldly sound and the 'jar' part of its name relates to this 'jarring' noise.

The Nightjar possesses a myriad of other folk names too:

Gabble Ratchet – may refer to the belief that the souls of unbaptised children were doomed to wander in Nightjar form until Judgement Day.

Lichfowl – meaning 'corpse bird', linked to the deathly superstitions surrounding it.

Fern Owl – reflective of its owl-like appearance and habit of roosting and nesting amongst ferns.

Dew Hawk – attributed to their pointy, hawk-like shape and their habit of becoming active at dusk as the dew forms.

Puckeridge – Nightjars were believed to strike calves with their beaks and infect them with a disease known as puckeridge, which became a name for the bird itself. The disease was instead caused by a fly that laid its eggs under the animal's skin.

STARS OF THE SCREEN

Given their secretive nature and preference for heathland landscapes, it took a long time before our *Springwatch* cameras managed to pry into one of these hidden nests. In 2022, we were able to bring a live Nightjar nest for the first time from Dersingham Bog in Norfolk, but it was our 2023 nest at RSPB Arne that delivered one of the most shocking moments in *Springwatch* history.

After watching our female bird successfully incubate and hatch her two eggs, our mouths were open in astonishment as she proceeded to swallow one of her chicks alive and whole. After speaking to Nightjar experts up and down the country, it's believed to be a behaviour never documented before. What drove the female to such a shocking act is a mystery and proved that this enigmatic bird certainly has many more secrets to uncover …

WHERE TO SEE: *Heaths, moors and recently felled or planted conifer plantations are their favoured spots across England, Wales and southern Scotland. Good populations can be found on the Dorset heaths around Poole Harbour, the Gwydir Forest in Eryri and Galloway Forest Park in Dumfries and Galloway.*

WHEN TO SEE: *Dusk is the time to enjoy the Nightjar's magic. Be in place before the light fades on a calm June night to experience the start of their unearthly chorus and catch a glimpse of their flight in the half-light.*

Swifts

Supremely evolved for life on the wing, Swifts take what it means to be a bird to a whole new level. They spend almost their entire lives in the sky. They catch their insect prey on the wing, they drink by skimming mouthfuls of water from ponds, lakes or rivers and they're even agile enough to align in the delicate act of mating whilst airborne.

ASLEEP IN THE SKY

Their greatest trick lies in their ability to sleep whilst they're flying. It seems an impossible feat, yet Swifts do the impossible every day. They're capable of 'unihemispheric sleep', whereby they can shut one half of their brain off at a time. On ascending into the night sky, they align themselves into the wind and enter their resting state. In this half-awake/half-asleep state, they're able to remain alert enough to adjust their position to remain safely in the sky.

One of the first recorded instances of this being witnessed by humans was from a French pilot during the First World War. On a reconnaissance mission near Vosges, he encountered a cloud of Swifts apparently floating motionless in the air, lit by the light of the full moon.

> *As we came to about 10,000 feet … we suddenly found ourselves among a strange flight of birds which seemed to be motionless, or at least showed no noticeable reaction. They were widely scattered and only a few yards below the aircraft, showing up against a white sea of cloud underneath.*

Unihemispheric sleep is something found across birds and something you can see for yourself without having to ascend to 10,000 feet. Head down to your local pond and pay attention to the ducks. Look for those sleeping, with their head tucked under their wings, and notice their eyes. In areas where they may be roosting in a group, those at the edge of the flock often have one eye open to be aware of danger, whilst those in the middle are able to rest both. Those sleeping at the water's edge often choose to close the eye facing the water so that they can stay vigilant to any danger or disturbance that might come from land.

THE BIRD WITH NO FEET

It's no surprise, given their mastery of the air, that their scientific name, *Apus*, was Latin for a bird with no feet. Nobody had ever seen a Swift land and it was thought that they lacked the appendages to do so. They seem likely to have been the inspiration for the mythical 'martlets' of medieval times. These birds were said to be born in the sky without feet and spend a life on the wing, perhaps symbolising a restless longing for adventure and learning. These would often adorn medieval heraldry and can be found on the coat of arms of historical leaders such as King Richard II.

Swifts do have feet, although their legs are so short that they're only capable of an ungainly shuffling walk. The only time they ever need their feet is during the breeding season when they land to make a nest and lay eggs.

SPEAK OF THE DEVIL

Swifts come tearing back into UK skies in May, announcing their presence with raucous screams that gave them the reputation of being the Devil's Bird. Others believed them to be the shrieking souls of the dead and their love of nesting in old churches did nothing to dispel the myth as they swirled around the spires, disappearing into the cracks as evening fell.

GETTING DOWN TO BUSINESS

Once they've returned from Africa, they waste no time in settling down to breed. Natural nest sites in cliff faces and old woodpecker holes are a rare sight nowadays and the vast majority of Swifts use holes in buildings. They're incredibly faithful to their nests too and will return year after year over their almost decade-long average lifespan.

Their nest is a collection of whatever they've been able to catch on the wing – feathers, paper, straw, fluffy seed heads and even butterfly wings have all been recorded. Here they lay two to three eggs and raise their chicks until they're ready to leap into the outside world.

FALL OR FLY

Fledging time for a Swift is no mean feat. If their first flights see them crash land to the ground, it can be a death sentence for an inexperienced Swift. Given their wings are so long and their nest spaces often so cramped, it makes it difficult for them to build their wing muscles through flapping as many other young birds do. Swiftlings have solved this dilemma by following a strict workout regime ahead of their big day. They repeatedly lift themselves off the floor using their wings, doing sets of press-ups to strengthen the core wing muscles that they'll need for flight.

Providing all goes smoothly, a young Swift's first flight is a remarkable thing. Most young birds would be quite happy for their first flight to carry them a couple of metres to the safety of the nearest bush, but not the Swift. Once it launches into the air, it begins a flight that will take longer than most small birds' lifetimes. Swifts only land when they reach breeding age, which can take 2 to 3 years.

CLOCKING UP THE MILES

Swifts are birds of extraordinary numbers. Scientists have recently shown that the fastest Swifts can travel 500 miles a day during their migration – the distance between Edinburgh and Paris – whilst one bird tagged in the UK only took 5 days to reach West Africa!

In their lifetime, it has been estimated that they travel an equivalent distance each year to five times around the Earth's equator and that they clock up well over a million miles in their lifetimes.

They're only a British bird for a snapshot of the year when they arrive in May and depart in early August. Their fleeting presence is one of the things that adds to their allure, an almost blink and you'll miss it marker of high summer. So, enjoy the Swift whilst you can; it is truly one of nature's finest evolutionary marvels.

WHERE TO SEE: *Although their numbers have halved in the last couple of decades, they can still be seen in many towns and cities. Look for them particularly around old buildings and churches, which possess the cracks they need to be able to get into for nesting.*

WHEN TO SEE: *Anytime between May and early August. By July, first-year birds scoping out nest sites for the following year have joined the breeding adults, increasing the sight and sound of the Swift screaming spectacle.*

Kingfishers

There's nothing like seeing a Kingfisher.

Their high-pitched call often tells of their incoming arrival. Don't blink. Suddenly a bolt of blue, as if from another dimension, hugging the surface of the water. Their sharp sound escapes from them like crackling energy. As quickly as they appear, they are gone, leaving scant time to even comprehend their apparition. Rarely does one get to enjoy a Kingfisher for long.

Kingfisher

Thankfully their colours make even the most fleeting glimpse possible. Their burnt orange breast contrasts with their trademark blue. It is like no other hue, a shimmering cloak of deep azure, with an intense electric streak down the middle that gives them that tell-tale flash down the river.

TRICK OF THE EYE

The extraordinary thing about Kingfisher blue is that it isn't *actually* blue. Blue is one of the most difficult colour pigments for nature to produce and so Kingfishers do not possess a single drop. Instead, it's the feathers' physical structure that creates their amazing blues, scattering the light in a phenomenon known as the 'Tyndall Effect'.

SEASIDE GETAWAY

Kingfishers are more widespread than you might think – found on most rivers, streams and lakes throughout Britain. Cold winters are tough for a bird that needs to dive into the water for its food, and they will often move towards coastal areas for the reliable open water that saltwater guarantees.

FEATHERED FABLES

It's unsurprising that a bird as magical as the Kingfisher appears in its fair share of stories and superstitions. It was said that their feathers would increase the beauty of women who wore them or that they had the power to protect the holder from lightning. One legend connected to the Noah's Ark story tells of how the Kingfisher got her colours:

> *When Noah sent out the dove from the Ark to find land, he also sent a Kingfisher out over the stormy waters. To avoid the worst of the weather, she flew towards the heavens, into the azure blue where her grey feathers took on the colours of the sky. However, she inadvertently flew too close to the rising sun, burning her breast with its fiery orange glow, and she had to dive into the sea below to extinguish the flames. By the time she had returned from the clouds, the ark had gone, and so the kingfisher still flies ceaselessly along the water's surface to this day, trying to refind it.*

In Greek mythology, Alcyone and Ceyx were a devoted couple deeply in love. Ceyx, the king of Thessaly, set out on a voyage across the sea. Alcyone, his wife, was distraught and prayed to the gods for his safe return. However, the sea god Poseidon caused a storm that sank Ceyx's ship. In her grief, Alcyone threw herself into the sea. The gods took pity on the couple and transformed them into kingfishers. According to the myth, the gods also granted them calm seas and a period of tranquillity during their nesting season, which came to be known as the 'halcyon days', a phrase that has now become used for times of serenity and peace.

BULLET BIRD

On the rare occasions that one does get to enjoy the sight of a Kingfisher for any length of time, you may be lucky enough to be watching one plunging into the water on the hunt for prey. So perfect is their body shape for piercing through the air that us humans looked to it for help with one of our own problems. The bullet trains of Japan were getting so fast that the build-up of air pressure on the front of the train was causing loud booming sounds when they would exit tunnels. Engineers noticed how seamlessly Kingfishers dived and decided to apply their aerodynamic shape to the front of the bullet train. Not only did it sort out the noise problem, but it also saved up to 15 per cent more energy due to increasing the trains' aerodynamic efficiency!

WHERE TO SEE: *Many wetland nature reserves have hides with sticks placed strategically in front of them to act as Kingfisher perches. It's just a case of having the patience to wait!*

WHEN TO SEE: *Although they can be seen all year round, winter is often a good time to spot them when riverside branches are bare. For a very colourful bird, it's amazing how they can disappear behind foliage!*

Ring-necked Parakeets

The London soundscape is a crescendo of noise – the traffic, the people – but there is also a bird that adds its own iconic sound to the chorus. Like the rest of the city's sounds, it is loud, even considered by some to be abrasive. Look up to greet the squawking sound and you will see its owner. With electric green feathers, a vivid red beak and a long, pointed tail, there is no chance of mistaking its identity. This is a British bird like no other: the Ring-necked Parakeet.

THE GREAT ESCAPE

London's parakeets have been famous for some time, although it was actually in Kent in 1969 that they were first recorded breeding. There are some fabulous rumours about how Britain's parrot population came to be. One is that a pair was released by legendary guitar hero Jimi Hendrix on Carnaby Street whilst visiting London in 1968. Another states that a flock of parakeets escaped from the set of the 1951 movie *The African Queen*, which was being filmed at Isleworth, near London. Another theorises that debris from a plane crash in the 1970s fell through an aviary roof in West London, setting free the parakeets held inside.

Ring-necked parakeet

However, the truth of the parakeet's arrival on these shores is far more mundane. As a popular pet, they've been kept up and down the country for a long time. Over the years, either through escapes or deliberate releases, birds have found themselves in the wild and managed to make a living for themselves.

MAKING THEMSELVES AT HOME

The most recent population estimates put them up in the tens of thousands and there are now well-established populations in major cities like Liverpool, Birmingham, Manchester, Newcastle and even as far north as Edinburgh and Glasgow.

HOW DO THEY SURVIVE HERE?

Genetic modelling has traced the origins of British parakeets to northerly parts of their native range in Pakistan and northern India, where they may be more adapted to deal with colder weather. They seem to rely on our cities and towns, with the relative warmth these bring compared to the surrounding countryside and the reliable feeding options that come with access to artificial food in bird feeders.

Ours represents the most northerly population of Ring-necked Parakeets anywhere in the world, but their colonisation of Britain isn't a one-off story. It's a trick they've repeated all across the world, with feral populations now existing in at least 34 countries across 5 continents.

FRIEND OR FOE?

The jury is still out on whether they cause any negative effects to native wildlife, with some evidence from the Continent that they may impact nesting birds and roosting bats by being able to outcompete them for nesting holes.

ALIEN VS PREDATOR

In winter, they can flock together in their thousands in the evening, coming together for safety in numbers to roost. Their emerald-green colouration makes them easily spotted by aerial predators and urban Peregrines have been shown to be quite partial to parakeets, so gathering together in huge numbers increases each bird's chance of safety.

These roosts can be quite the spectacle of sight and sound, and with the parakeet's rampant success, it looks like something we'll surely be seeing more of up and down the UK.

WHERE TO SEE: *There are many areas in England now where parakeets are firmly established. Good spots to see them include St. James's Park in London, Sandwell Valley in West Bromwich and Walker Park in Newcastle.*

WHEN TO SEE: *These birds are resident and obvious at any time of year, but their winter flocks are quite a sight (and sound!) to behold. Roosts of thousands can be found in London and areas of Kent where parakeet populations are highest.*

Woodpeckers

If there's one bird sound we can surely all recognise, it's the characteristic drumming of a woodpecker. In Norse mythology, the sound was likened to the cracking of thunder, and woodpeckers became associated with Thor, the god of thunder and lightning. While Thor wielded his magic hammer Mjölnir, it is the woodpecker's strong, sharp beak that is its tool of choice.

Woodpeckers were said to be the key to finding a plant with legendary powers. Springwort was a mythical herb believed to draw down lightning and it was said that woodpeckers would eat it at midnight on Midsummer's Eve to gain the strength needed to hammer through the hardest of oaks. Springwort was capable of opening any closed or locked door but was impossible for any human to find. One recorded technique of obtaining springwort was to block up the hole of a nesting woodpecker. The woodpecker would then seek out a sprig of springwort and bring it to the nest, causing the blockage to instantly be dislodged.

Wrynecks

Once a bird found nesting throughout the UK, the Wryneck was lost as a breeding species when the last ones disappeared from Scotland in the early 2000s. A loss of suitable habitat has seen the Wryneck fade from Britain as a reliable presence throughout the summer, but it continues to delight birders as it passes through in spring and autumn on migration.

CRYPTIC COAT

This is a woodpecker unlike any other, dressed in a remarkable cloak of mottles and streaks in every imaginable hue of browns and greys. It is only a small bird and this plumage gives them the ability to melt away against bark and bare earth. They feed almost exclusively on ants, which they forage from the ground, and are highly migratory, unlike our other woodpeckers, spending their winters in sub-Saharan Africa.

DO THE TWIST

Their most notable feature gives them their name. When threatened, Wrynecks move their head in a strange 180-degree figure of contortion, believed to possibly be an attempt to mimic a snake to ward off predators. In some parts, 'snake bird' was used to describe it and this unnerving behaviour led the Greeks to believe it had magical powers and could be used as a love charm.

WHAT'S IN A NAME?

Their scientific name, *Jynx torquilla*, is derived from the Greek story of Iynx, a Greek nymph that was turned into a Wryneck by Hera after she discovered Iynx had made her husband Zeus fall in love with one of her priestesses. This story gives us the first part of their scientific name, and the word jinx – a curse or bad luck spell. The second part of its scientific name – *torquilla* – translates to 'little twister.'

WHERE TO SEE: *Your best chances lie with bird migration hotspots on the eastern and southern coast. Places like Portland Bird Observatory in Dorset or Spurn Bird Observatory on the Yorkshire coast have them passing through most years.*

WHEN TO SEE: *Although they can be seen in spring too, their autumn passage in August to September is when Wrynecks appear on British shores in the greatest number.*

Great-spotted Woodpeckers

The Great-spotted Woodpecker is a species going from strength to strength and is the bird that comes to mind when many of us think of woodpeckers. Their striking colour scheme is enhanced by a flash of vibrant red under the tail and also on the nape of male birds.

ON THE UP

Across most of Britain, if there are trees, then you stand a pretty good chance of seeing a Great-spot. They've seen an astonishing population increase of 403 per cent in the last 50 years, recently extending their range further into northern Scotland and colonising Ireland, where there is a small but quickly growing population.

HEADBANGERS

Their short, sharp bouts of drumming are often heard ricocheting around woodlands in early spring, well before the trees have donned their leaves. This is their version of a song, which they use to attract mates and deter rivals. The drumming of a woodpecker is an astonishing thing and they can strike the wood up to 20 times a second to produce the sound. Their heads travel at speeds of up to 7 metres a second and, on impact with the wood, come to a complete stop. This generates an eye-watering force of more than three times the human concussion threshold. Just how they manage to avoid brain injury is still contested. Originally, it was thought that the skull of the woodpecker was soft and able to act as a sponge to absorb the force of the impact, but recent studies have contested this. Their small size means that they're subject to a lot less force

DID YOU KNOW?
Woodpeckers want the loudest drumming, and so will search out the perfect branches with a resonance that will carry their sound the furthest. It's not just trees they'll use to stake their territorial claim either: they'll often use metal – particularly pylons!

than we would be in similar circumstances, and the way their brain is positioned and packed tightly into the skull reduces their risk of concussion.

ON THE MENU

The sound of a woodpecker's drum is often misidentified as the sound of them drilling holes, but that's a much more rarely heard methodical, slow hacking into the wood. This is also how they hunt for their main source of food – insects hidden under the bark. Part of their recent success is because of their catholic tastes and, when it comes to food, there isn't much that a Great-spotted Woodpecker would turn its nose up at. They're the species of woodpecker you almost always see on garden feeders, helping themselves to fat balls, peanuts and suet pellets.

STARS OF THE SCREEN

Throughout the years on *Springwatch,* the Great-spotted Woodpecker has found itself cast as one of the programme's villains. They will regularly seek out the nestlings of other birds, particularly if they have their own chicks in the nest to feed. They've been caught in the act multiple times by our *Springwatch* nest cameras, peering ominously into Blue Tit boxes or successfully raiding Treecreeper nests and carrying away the chicks.

In *Springwatch* 2019, they were given their own starring role when a small camera fitted *inside* a Great-spotted Woodpecker nest in Abernethy Forest gave us an extraordinary view of their secret lives. We watched as the chicks grew in strength to climb up to the nest hole, fighting each other for the prime position to be fed by the adults. With another camera at a feeding station elsewhere in the wood, we were able to see the male woodpecker collecting food and flying it back to the chicks. Through timing how long it took him to leave the feeding station and arrive back at the nest, we were able to work out a brand-new statistic for science – that the air speed velocity of a laden Great-spotted Woodpecker is 8.3 metres per second!

WHERE TO SEE: *Look for Great-spotted Woodpeckers in woodlands, listening out for their short, sharp 'kick!' call as a sign of their presence in the treetops. Feeding stations give you the best chance to enjoy them.*

WHEN TO SEE: *They will begin their territorial drumming in winter and continue right through into spring. This is often the best time to locate them before the trees get their leaves. Their chicks are also incredibly noisy in the nest, so listen out for their loud calling in late spring.*

Lesser-spotted woodpecker

Lesser-spotted Woodpeckers

The Lesser-spotted Woodpecker is no bigger than a House Sparrow and gives the impression of a Russian doll you'd find inside a Great-spotted Woodpecker. Given their size, they are notoriously hard to find. Add to this their decline of 72 per cent since the seventies, and there's a reason they're called Lesser-spotted!

TINY BIRD WITH A TITANIC TERRITORY

They rely on standing dead wood for much of their food, foraging for the invertebrates that live on it. To make sure that their territories encompass enough standing dead trees to raise their families, they need to make them large. Some estimates have it that one pair of these birds can require as much woodland as 500 pairs of Blue Tits! Sadly, as we fragmented our woodlands and tidied up the bits that were left, standing dead wood became a rarer commodity and so Lesser-spots became scarcer.

LESSER SPOTTED BY NATURE ...

Outside of the courtship season, this little bird can be seemingly impossible to find. Once their territories are established, they fall largely silent for another year, melting back into the forest and going about their lives mostly unseen. In autumn and winter, they will join mixed species flocks and roam woodlands with similar-sized birds such as tits and nuthatches. Teaming up is a good way of being able to find more food during the colder months and more eyes mean more chance of spotting any predators!

WHERE TO SEE: *Their populations have retreated to larger areas of good habitat with enough standing dead trees. Spots like the New Forest in Hampshire, Sherwood Forest in Nottinghamshire and Blean Woods in Kent still hold resident populations.*

WHEN TO SEE: *The best hope of finding these secretive tiny birds is in late February and early March, when they drum on trees to attract mates and defend territories. Listen out for a longer, flatter drumming sequence than a Great-spotted Woodpecker and you might catch a glimpse.*

Green Woodpeckers

Green Woodpeckers didn't get the same dress code memo as our other two resident woodpeckers. In fact, they couldn't have come wearing a more outlandish outfit if they tried. With their green body, bright red cap, yellow rump and bold, black face, they look more like a children's entertainer when compared to the sharply dressed monochrome of their Lesser- and Great-spotted cousins. Their gaudy colours, coupled with their size, make them one of the most unmistakable birds in Britain.

WEATHER WARNING

It also possesses a call just as outlandish as its appearance. Known as a yaffle, its call is a far-carrying, laughing sound that has given it English folk names such as Yaffle, Yaffingale and Laughing Betsey. Its yaffling was also said to summon or tell of oncoming rain – leading to such names as Rain-bird, Rainpie, Rain Fowl, Weather Cock and Wet Bird. John Aubrey, writing in the seventeenth century, noted that 'to this day the country people do divine of raine by their cry'.

Green woodpecker

SPECIFIC TASTES

Green Woodpeckers favour open ground more than our other species, so look for them on heathlands, parklands and grasslands with scattered trees. It's here that they're able to find an abundance of their favourite prey – ants – which they feed on from the ground. Chicks need to eat around 1.5 million ants and their pupae before fledging!

STARS OF THE SCREEN

It was a Green Woodpecker nest at RSPB Minsmere on the 2016 series of *Springwatch* that provided one of the most dramatic instances of predation ever filmed by the programme when a stoat was filmed climbing 8 metres vertically up an oak tree to take the chicks. The stoat in question was a mother that had been filmed over the course of the series, hunting rabbits and carrying her young around. It was a sad end for the Green Woodpeckers, but the chicks provided a valuable meal to the young stoats.

WHERE TO SEE: *Green Woodpeckers are found across much of Britain, other than the far north of Scotland and Northern Ireland. It is worth checking any sites where there is open land with trees for them to nest in.*

WHEN TO SEE: *Any time of year. They are more vocal in the spring, and their unmistakable yaffling makes them much easier to find.*

Skylarks

Skylarks are birds of open landscapes with big skies. Grassland plains provide their natural homes, but our farmland fields and managed moorlands tick all the boxes required for them to thrive too. Sadly, as farmland has been worked ever more intensively, Skylarks are among our most declining bird species.

MEET THE COUSINS

The population of our other breeding Lark species, the Woodlark, is thankfully moving in a much healthier direction. This is a rarer bird, more restricted in its range than the Skylark and associated with wooded heaths and edges. It possesses a bold white stripe above the eye and a more lilting song than its cousin, given in display flights consisting of mesmerising wide circular arcs rather than vertically ascending. Monitoring over the last few years has seen their populations recover well, and in 2015 they completed their transition from being a red-listed bird to being added to the green list as a reflection of the state of their numbers.

EARLY RISERS

If you are up and out at daybreak, then you are said to be 'up with the Lark' because of their habit of singing at dawn. The Lark has been used to reference the early hours for hundreds of years, with Chaucer referring to 'the bisy larke, messager of day' in *The Knight's Tale*, whilst Shakespeare's 'Sonnet 29' includes the words 'the lark at break of day arising / From sullen earth, sings hymns at heaven's gate.'

MUSICAL MASTERPIECE

It is the Skylark's song that has enshrined its place in human hearts throughout the ages. To us, it is a blur of rapid-fire notes unspooling from the bird in an epic, continuous length. It is incredibly complex, built from hundreds of syllables delivered at a rate of up to 43 per second. It is said that a well-tuned ear can identify whether the singing bird is ascending or descending on its song flight, as the song takes on a slower, quieter, minor key during the descent.

An old country fable goes that, on its ascending flight, the Skylark is joyous at the thought of entering Heaven, but having been refused entry by St. Peter, its song takes on a more melancholic tone as it returns to Earth.

GROUNDED

When they're not in the sky, Skylarks forage for food and nest on the ground. The newly hatched chicks of both species are adorned in extraordinary grass-like down feathers to camouflage them. Larks have unusually long hind claws for songbirds, an attribute that has evolved to aid their terrestrial life. This curious trait inspires a poem that claims to be a translation of the Skylark's song as it rises into the sky:

Up in the lift go we,
Tehee, tehee, tehee, tehee!
There's not a shoemaker on the Earth
Can make a shoe to me, to me!
Why so, why so, why so?
Because my heel is as long as my toe.

DID YOU KNOW?
A group of Skylarks are known as an 'exaltation', a word to describe a feeling of extreme happiness. It's a collective noun that can trace its usage back to around 1430 and is evidence of just how long the rapturous rising song of a Skylark lifting into the great blue sky has filled our hearts with joy.

CUSTOMS AND CURSES

Eat three lark's eggs on a Sunday morning before the church bells ring and it was said you would be bestowed with a lovely voice. However, in Shetland, the Skylark was treated with great respect and the birds and their nests were not to be disturbed. There are three black spots upon their tongue, each one representing a curse that

Skylark

would befall you if you were to harm them. Tongue spots can actually be found on many species of birds, particularly when they're chicks in the nest. It's thought that these markers in the middle of their mouths help the parents to aim the food more successfully into the gapes of their hungry young.

EASTER ORIGINS

One story attempting to explain how the Easter Bunny came to be centres around the Skylark. It is said that Ēostre, the pagan fertility goddess of humans and crops, finds a wounded Skylark frozen on the ground. To ensure it can survive the winter, she transforms it into a hare. The hare found it could still lay eggs, so it decorated these each spring and left them as an offering to the goddess.

There's a logical explanation for this tale. Hares make depressions in the ground amongst the grass known as forms. When the hares move on, these empty forms can be taken over by Skylarks to use to make their nests. It may have been that farmers saw eggs lying in a hare's form and concluded they must have been laid by the hare itself, thus spawning the legend.

STARS OF THE SCREEN

Nesting on the ground is a dangerous place for any bird, but one predator we didn't expect to see snuffled onto our screens during *Springwatch* 2022 at Wild Ken Hill. Having gone 17 years without ever recording a hedgehog on one of our live cameras, the moment one flushed a female Skylark off her nest in the dead of night and snaffled up the chicks was certainly something we didn't see coming!

WHERE TO SEE: *Skylarks can be found in flat, open habitat of many different types, such as farmland, grassland, saltmarsh and moorland. Listen for their song and look skyward, searching for that little black dot in the sky! Woodlarks are best looked for on the heaths of East Yorkshire, East Anglia, the New Forest and Dartmoor.*

WHEN TO SEE: *It's best to look for both species on warm, calm mornings of early spring when their song helps to track them down.*

Hirundines

Hirundine is the name given to the family of birds that includes the Martins and Swallows – derived from the Latin word for the latter. Swifts, although similar in body shape, are a separate family of birds altogether and their common ancestors diverged millions of years ago. One of the key differences is that the Hirundines have the ability to perch like songbirds. They have three toes forward and one toe back, in an arrangement known as anisodactyl. Swifts have all four toes pointing forward, known as pamprodactyl, which allows them to grip onto vertical surfaces but makes it impossible for them to perch.

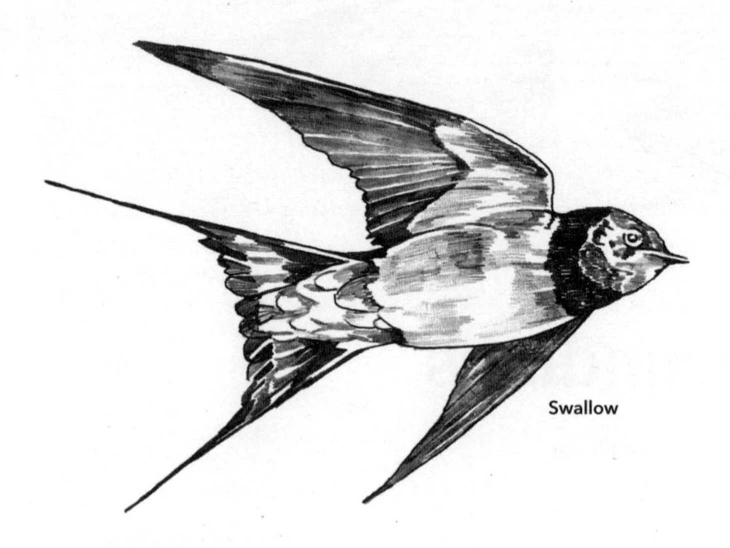

Swallow

Swallows

Skimming over the tops of fields with unparalleled acrobatic ease, the Swallow is *the* sign of summer to lift our spirits in April.

A FAMILIAR FAVOURITE

They've held onto this position in our hearts because of the proximity they can be found to us. Swallows were one of the great winners when humans cleared Britain of much of its forests thousands of years ago and began farming on a larger scale. To a Swallow, these open landscapes full of livestock were perfect, and the barns we built to house these animals made for a seamless, and much more abundant, replacement for the caves to which they had been previously confined.

And so it was that they became one of our most familiar birds, living alongside us and working their way into our hearts, the Swallow's arrival each year leading us out of the dark gloom of winter with a promise of better times ahead.

FEATHERED FOLKLORE

The Swallow is a lucky bird and, throughout Europe, killing one or destroying their nest has been counted as a sin. Within a year of committing any crime against the Swallow, expect your cows to yield bloodstained milk, for it to rain for a month or for there to be a dreadful misfortune to someone in the family. It was also said that watching their behaviour could help predict the weather, with rhymes such as:

Swallows fly high, no rain in the sky;
Swallows fly low, 'tis likely to blow.

NO PLACE LIKE HOME

Swallows make their nests from mud, gathering it in their beaks from puddles in farmyards before mixing it with their saliva and moulding it into a basic platform on top of a beam or other such structure. They then line it with a sprinkling of feathers before laying four to five eggs that they will incubate for around 17 days.

A Swallow nest in a building was said to protect it from lightning strikes, fire and other evil mishaps. In Scotland, a Swallow nest on the window brought prosperity, whilst in Somerset, it set those who lived in the house up for health, happiness and wealth. It was a bad omen if the birds did not return the following year. In Normandy, it was said that Swallows were able to find special stones from the seashore that could restore sight to the blind, which they tucked away in their nests.

STARS OF THE SCREEN

Swallow nests have featured numerous times on *Springwatch* over the years. Usually everything progresses happily, with Swallow nests seemingly more sheltered from predators because they are tucked away in buildings. However, one of the most extraordinary observations from the *Springwatch* cameras came in 2008 in Pensthorpe, in Norfolk, when a male Swallow was observed landing on the edge of the nest, only to pick up the young chicks and drop them to the floor.

The best explanation for this behaviour was that this male had either paired with, or was planning to pair with, another male's female. It wouldn't have been in the interest of our murdering male to sit by and watch this female raise another male's chicks, so he was eliminating them in order to make her ready to mate again and turn her focus onto chicks with his own DNA in them.

It was certainly a reminder that nature can act in some grizzly ways in even our most cherished species – and this was the first time this behaviour had ever been documented in Swallows!

HEADING SOUTH

Once Swallows have finished breeding at the end of summer, large numbers can often be seen grouped together on telegraph wires as they prepare for their long journey south. In Norfolk, there was a belief that these babbling gangs of Swallows that clustered together were discussing who amongst the human residents would die by the time they returned from their journey south next spring.

Another of their favourite places to gather before heading off on migration is reedbeds, where huge numbers amass for safety in numbers. Their sudden disappearance following these gatherings led some to believe that the birds must have dived underwater to hibernate. They were said to join their beaks, wings and feet together and stay there until the warm weather returned the following spring. There are even sixteenth-century drawings of fishermen pulling up net-loads of hibernating Swallows from a lake.

DID YOU KNOW?
Thanks to our warming climate, it has been possible to find Swallows in Britain throughout the year, even in the depths of winter. Temperatures have stayed warm enough in areas of southern England in recent years that Swallows can continue catching the flying insects they need to survive right through until spring.

The truth of their epic journey was only revealed in the early twentieth century thanks to bird ringing. In December 1912, a Swallow was caught on a farm in Natal, South Africa, with a ring on its leg that had been fitted 18 months earlier by John Masefield, an amateur naturalist, on the porch of his home in Staffordshire.

WHERE TO SEE: *Swallows are found right across the British Isles. Look for them skimming low over fields, particularly around livestock or over meadows, where there are plenty of flying insects.*

WHEN TO SEE: *They arrive in April and hang around into October. A late summer roost in a reedbed can attract hundreds of birds before they leave on their migration.*

House Martins

All dressed up in its finest, the black-and-white tuxedo colour scheme of a House Martin makes it a dapper little bird. On closer inspection, the black is actually rather a glossy blue, a deep, shimmering inky hue that's almost impossible to appreciate in the field. In Somerset, it was once believed that the Swallow and the Martin were husband and wife.

KEEPING WARM

Another feature unique to House Martins is their very fluffy legs. Why exactly House Martins seem to need to keep their legs warmer than their cousins isn't exactly clear but may be down to where they catch their prey. House Martins generally hunt for insects at a higher altitude than Swallows, where the air is cooler so any exposed skin would result in a lot of heat loss. Furthermore, there is some suggestion that House Martins may even pull the same trick as Swifts when they're in their African wintering grounds and sleep in the sky, meaning their fluffy legs would keep them warm!

MUD MARVELS

House Martins are best known, for better or worse, for how and where they choose to nest. The 'where' is self-explanatory and the vast majority of House Martins rely on buildings to use as nest sites. House Martins didn't just sit around idly waiting for humans to build houses, however, and there are still sites around the British coastline where you can see them nesting on their ancestral home of steep cliff faces.

House Martins have one of the most recognisable nests of any bird in Britain. They use the same materials as Swallows – mud and saliva – but House Martins take their craft to a whole new level. Rather than making a simple platform structure on top of an existing base, House Martins will create an entirely domed structure built up and out from the walls of a building to connect to the eaves. There will be a small entrance hole just big enough for the bird, either on one side or at the front, and the inside will be stuffed with soft feathers. House Martins like to live colonially, so sometimes join their nests together like a row of terraced houses. These colonies are wonderful places to be, filled with the telltale sound of chattering House Martins as the birds zip backwards and forwards to their nests.

SHARING OUR SPACES

These nests aren't always welcomed. The issue comes when the chicks, in an attempt to keep their nest clean, eject their droppings all over the floor below. If the nest is under the eaves of a house, then this can lead to quite a messy situation, which isn't always the most appealing to live with. However, with a population decline of over a third in the last 25 years, any help we can give the House Martin is vital. An easy fix is to fit a tray beneath the nest site to catch the droppings before they hit the ground.

WHERE TO SEE: *Look for them disappearing into their nests under the eaves of houses. They like to nest in colonies, so houses or streets with one nest will often feature multiple. They can also be seen on muddy riverbanks or puddles, collecting the mud they need to build their nests.*

WHEN TO SEE: *They arrive in Britain in April and have left again for Africa by October.*

House martin

Sand Martins

Easily confused with the House Martin is the Sand Martin. The main difference in their appearance is their colour, with the blacky-blue of House Martins being replaced with a brown colour more reminiscent of, well, sand. There are two key patterning differences too. Sand Martins are all brown on top, without the white rump of their cousins. Underneath the role is reversed, and rather than the clean, pure white belly of a House Martin, Sand Martins have a brown strap around their chin, as though they're wearing a bicycle helmet.

RIVERSIDE APARTMENT

They are one of our earliest returning migrants and can be seen flitting low over water bodies looking for insects on chilly March days. They have a great fondness for water given that it provides the type of habitat they need to nest in – sandbanks. Sand Martins make their nests at the end of long tunnels that they dig out into vertical sand along the edges of rivers, lakes, quarries and the like. The nests can stretch almost a metre back and there can be so many in one bank that the pockmarked holes resemble a slice of Swiss cheese. Sand Martins are easily encouraged to nest in artificial banks and there are many nature reserves around the country that have specially built structures next to water bodies for Sand Martins to nest in.

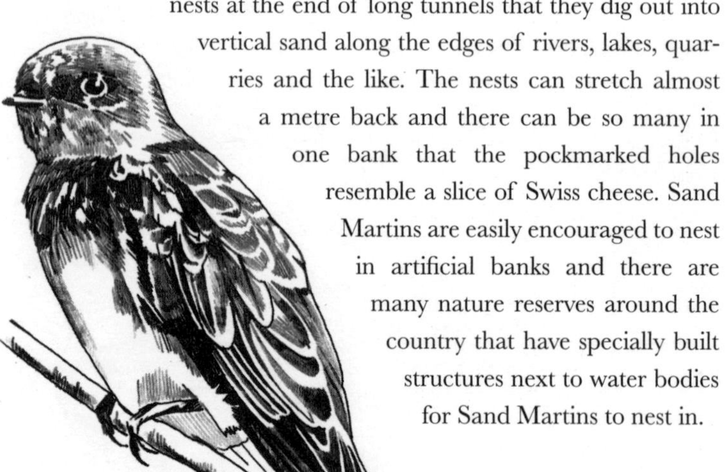

Sand martin

WHERE TO SEE: *For the best views of Sand Martins, look for nature reserves that have specially built Sand Martin banks for them to nest in. There are many around the country, such as Attenborough Nature Reserve in Nottinghamshire, Gailes Marsh in North Ayrshire and the London Wetland Centre.*

WHEN TO SEE: *They can be seen right throughout spring and summer but, as one of the first African migrants to return to Britain in early March, spotting them at this time of year means they're less likely to be confused with their closely related cousins that are still in warmer climes.*

Dippers

Never has a bird been more aptly named than the Dipper. Perched on a rock beside a fast-flowing stream, they are famous for their fidgety bobbing motion, which occurs almost constantly whenever the bird stands in one spot. Watch one for long enough and you will see it suddenly dive below the water, emerging back onto a rock with an invertebrate foraged from the stream bed.

They are compact little birds, with a dark plumage over most of the body contrasting with their brilliant white chest. Their appearance led to them being once classified as a member of the thrush family and names like Water Ouzel, Water Thrush, Water Blackbird and Brook Ouzel have all been used in the past.

Dippers are their own family, made up of just five species spread across the globe, that are specialised in living in watery habitats. They are unique amongst songbirds for being the only ones who live an aquatic lifestyle and having the ability to dive under and swim beneath the surface of the water.

WHY DIP?

Why exactly Dippers dip is still a mystery, although there have been multiple theories put forward to explain their incessant bobbing. One suggests that the dipping behaviour, against a backdrop of rushing, turbulent water, helps to conceal the bird from predators by making its shape harder to pick out. Another theory has it that by constantly moving, the birds are better able to see any prey beneath the fast-flowing surface.

However, it is the third theory that seemingly holds most weight. The environment that Dippers live in is often noisy and the constant sound of water makes communication between Dippers rather difficult. They have loud, high-pitched calls, which help to cut through the sound of the torrents, but that only goes so far. It is thought then that the dipping motion is a way of talking to other Dippers without the need to shout above the noise. Dippers also have white eyelids, which they rhythmically close as they bob, adding support to the theory that the birds are trying to make themselves seen to members of the same species.

WHERE TO SEE: *Resident Dippers are only found in northern and western Britain and over much of Ireland, where the right kinds of fast-flowing, shallow streams exist for them to live along. Keep scanning the rocks as you walk along a stretch of suitable river and remain patient as their territories can extend for some distance up and down the waterway!*

WHEN TO SEE: *Dippers are resident all year round and will often nest under bridges over streams in spring, meaning that adults feeding chicks can be easily viewable.*

Dipper

Wrens

The Wren is an unassuming little bird with a big reputation.

In an old fable, it was crowned the 'king of birds', when a contest amongst them was arranged to see who could fly highest. Whoever did so would be given the royal crown and soon it was the eagle soaring above the rest of the competition. The eagle was certain it had claimed its crown, only for a Wren, which had stowed away on the eagle's back, to emerge and fly higher. So enraged was the eagle, that it flew to the mountains to live out its days in solitude, whilst the other birds acclaimed the Wren to be king thanks to its cunning.

USURPER TO THE THRONE

It may be that the Wren has stolen its kingly title from another bird too. Goldcrests, even daintier and somehow more minuscule than Wrens, used to be known as Golden-crested Wrens. Their Latin name, *Regulus*, means 'little king' and its name in languages like Swedish *(kungsfågel)* translates to 'Kingbird'. It would seem more fitting for the bird with the golden crown to be deemed king, but it seems the Wren has pulled off yet another royal heist in stealing this name for itself!

HIDDEN ALL AROUND US

Wrens are incredibly skulking, hopping around low in the undergrowth, appearing more like little mice when you glimpse them between the brambles. When they reach a gap in a hedgerow, their flight is a frantic, whirring blur of short brown wings, like a clockwork toy wound up furiously until its release. If you do get a clear view, that cocked tail standing proudly is a telltale sign of the Jenny-Wren.

Wrens' scruffy, domed nests are stuffed into ivy, gaps in walls, garden sheds, under bridges or anywhere else they deem suitable.

Their scientific name, *Troglodytes troglodytes*, means 'cave dweller', and it is true that a Wren's nest can often be found jammed into the cracks around cave entrances.

LITTLE BIRD, BIG VOICE

You're far more likely to hear them than see them. For their size, they have an incredibly loud song and will perform it for most of the year. Its rapid-fire, high-pitched ringing notes carry far and will often set off a chain reaction of other male Wrens calling back in defence of their own territories. Key into their song and you'll learn there are many more around than you ever see!

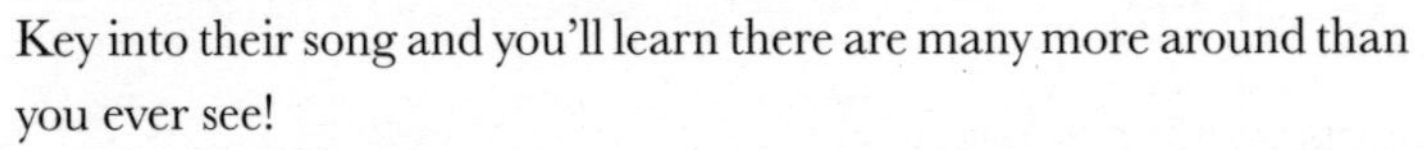

Wren

CURIOUS CUSTOMS

The Wren enjoys a curious relationship with British folklore. Anyone who dealt harm to them was to be delivered serious misfortune in the form of a curse on themselves or their livestock. That being said, there is a notable exception to the rule. On either St. Stephen's Day (26 December) or Twelfth Night (5 or 6 January), the hunting of the Wren would bring fortune and prosperity for the coming year.

Wrens would be captured, killed, placed in a box or displayed on a pole and paraded with song from door to door by 'Wren Boys', who would seek food and drink from the homeowners, presenting them with a feather if they obliged. The Wren Hunt still continues in some parts of the British Isles, although the need to use a dead Wren has thankfully been replaced with more humane substitutes.

DID YOU KNOW?
The Wren is actually Britain's most common bird. The most recent population estimate puts them at around 11 million pairs!

WHERE TO SEE: *Wrens can be found almost anywhere – woodland, scrubland, parks, gardens, even open moorland and coastal cliffs. Wherever there are nooks and crannies for them to hide their nests and cover for them to find food, there will be Wrens!*

WHEN TO SEE: *Wrens will sing all year round but become much more vocal in spring. You may be lucky to catch a glimpse of one singing from an exposed spot, but mostly they sing from cover. Narrow down the bush where the sound is coming from and you may be rewarded with a glimpse as it skulks away.*

Waxwings

It may seem odd for any exotic-looking bird to arrive in Britain when the days are darkest and the frost bites at your fingers. They don't grace us with their presence each year, at least not in big numbers, but if the conditions are right, then Waxwings will leave their homes in the northern forests of Scandinavia and Russia to seek out food.

Waxwings are exquisite birds that give off an air of being handcrafted. Slightly smaller than a Starling, their plumage is made up of soft greys and browns, with vibrant accents added like fine details on a painting. Their crest is their most notable feature, long and neat on the top of their head, and can be raised or lowered depending on their mood.

WHAT'S IN A NAME?

The Waxwing gets its name from the red tips on the end of its flight feathers, which look like they've had the ends dipped in sealing wax. The reason for these unusual feather adornments is still not fully understood but it may help in mate attraction.

Of the three species of Waxwing in the world, it's the Bohemian Waxwing that visits our shores. The 'bohemian' part of its name may refer to the bird's wandering nature, and it can be difficult to predict when we might be in for a 'Waxwing winter'.

Waxwing

BERRIES

Despite looking like intricate ornaments that must be fed on a diet of gold and precious jewels, they are nothing more than highly evolved berry-scoffing machines. They will drift across the lands until they find a good berry crop to station themselves on for a while, gorging on them until they're gone.

If the berry crop is poor, or the Waxwings have had a good breeding year and produced a lot of new hungry mouths, then they are required to travel further to find what they need. This is when the Waxwings arrive in big numbers, in events dramatically called 'irruptions'.

HAUNTS

Irruption winters are unpredictable given the conditions required and can result in several thousand birds arriving in Britain. This is your chance to see them and, luckily for any wannabe waxwing watchers, this magical bird has a habit of hanging around in very un-magical places. Their search for berries regularly takes them into towns and cities, where they have no problem hanging out around people.

One of their favourite haunts is the glamorous backdrop of supermarket car parks, where their distinctive trilling calls often give away their presence as they feast on rowan berries. It's often not hard to find Waxwings though. Word will soon get out amongst birders when there's a flock of Waxwings in town and soon the car park may be filled with as many people carrying cameras and binoculars as it is people wheeling trolleys!

So, make sure to experience them whilst you can, before the berries disappear and they continue their wandering. Their journey will eventually take them back to the northern forests, with no telling when they might next return and bless us with another opportunity to enjoy them …

WHERE TO SEE: *Waxwings are unpredictable and can turn up anywhere in good years. Your best chances are further to the east of Britain though, where they will make landfall first after crossing the North Sea. During irruption winters, any rowan berry-laden trees are worth checking!*

WHEN TO SEE: *They arrive as winter sets in, from late October, and will roam across Britain throughout the colder months.*

Robins

In 2015, the result of a vote to crown the National Bird of the UK shocked absolutely nobody when the Robin was announced the victor. With 34 per cent of all ballots, the Robin dominated the competition, trouncing the second-place Barn Owl and third-place Blackbird, with 12 per cent and 11 per cent respectively.

THE NATION'S FAVOURITE

The British have a unique relationship with Robins. There have always been superstitions around what bad fortune you would encounter if you were to harm a Robin. Uncontrollable shaking, sickness and broken limbs are all a risk, whilst in Wales it was said that destroying their nests could bring a death to the family or a fire or lightning strike to the house.

In a time when it was open season to capture or kill many birds, the Robin seems to have had an air of unofficial protection for centuries. Despite its beautiful song and the practice of keeping other songbird species in captivity, the Robin seemed to largely escape this treatment. William Blake wrote in 1803:

A Robin Red breast in a Cage
Puts all Heaven in a Rage.

Our love for Robins owes itself to their bright, bold appearance and confident behaviour. They're known all over the country for their habit of following gardeners around as they turn over the soil, a trait that they've transferred to humans after thousands of years of following wild boar around as they rootle through the ground. Their desire to seemingly be close to us gives them a friendly appearance and there is many a Robin that has become tame enough to be handfed.

ROBIN RED BREAST

So, how did the Robin get its most famous feature? In some stories it is said that the Robin plucked the thorns from Jesus Christ's torturing crown to ease his suffering, and it was the blood of the Messiah that spilt upon the bird's chest. In some cultures, the Robin is a 'fire bird', one of the birds used to explain how fire came to Earth. It was in carrying the flames that the Robin scorched its breast, giving it the warm orange glow that it possesses today.

Its iconic colour has been used as a name for centuries. The Anglo-Saxons called it *rudduc* in reference to its ruddy colour but it is Robin Redbreast that is the most famous. This was actually the bird's full name in the early 1500s and was then shortened to Robinet or Robeen before it became Redbreast in the nineteenth century. It was only in 1952 that the British Ornithologists' Union formally accepted Robin as an alternative to Redbreast.

EVERY BIRD FOR THEMSELVES

The Robin's song is a high-pitched, thin, silvery song, often trailing off into fading, whimsical whistles. As autumn arrives, male and female birds have gone their separate ways following the end of the breeding season. Now they set up their own territories, which they will defend right through until it is time to pair up again next spring.

This need to protect their own patch throughout the leaner months leads females to sing too. This is a rarity amongst British birds and, whilst it is only males that sing in the spring, a Robin heard singing in winter could be either sex. The winter song of a Robin is an uplifting sound on dark, dreary days and can often be heard under streetlights in the dead of night.

FESTIVE FEATHERED FRIENDS

Robins are most associated with Christmas and regularly feature on greetings cards with their bright orange breasts beaming out against a snowy festive scene. The root of this association comes from the Victorian times when Royal Mail postmen would wear bright red uniforms, which earned them the nickname 'Robins'. As the 'Robins' would be the ones delivering Christmas cards, it wasn't much of a leap to start depicting the actual bird on the cards themselves, often sat upon a post box.

WHERE TO SEE: *Robins are everywhere and are not difficult to find given that they will often come to you! In some areas around nature reserves where people have fed them regularly, tame Robins will hop onto your hand for a snack.*

WHEN TO SEE: *Winter is often when they're at their most obvious and it's the perfect time to learn the Robin's song. When the rest of the chorus has fallen silent until the spring, the Robin can be heard clearly, trilling its merry little song into the cold air.*

Nightingales

'Thou wast not born for death, immortal Bird!' wrote John Keats in his 1819 poem 'Ode to a Nightingale', inspired by the everlasting quality of a song that lives long beyond the lifespan of the bird singing it. It is said that the poem was written when a Nightingale nested that spring near to the house that Keats shared with his friend, Charles Armitage Brown. Brown wrote: 'Keats felt a tranquil and continual joy in her song; and one morning he took his chair from the breakfast-table to the grass-plot under a plum-tree, where he sat for 2 or 3 hours.'

It was there, listening to the Nightingale's powerful repertoire of notes, that Keats composed his piece. Of course, it would have been the male bird of the pair doing the singing, as it is their famous chorus that has captured the hearts of poets and songwriters. Males will sing throughout the night until they have found a mate, after which their singing becomes restricted to dawn and dusk.

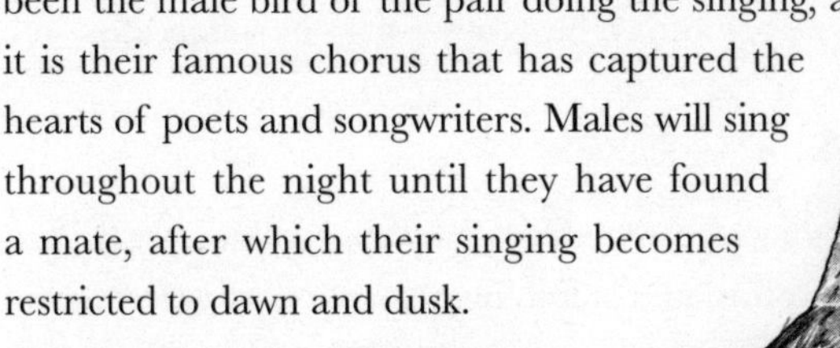

TURNING DOWN THE VOLUME

Sadly, Keats' vision of the Nightingale's enduring immortality has not come to pass. Over the last 25 years, the UK breeding population has declined by a half and it can now only be found reliably in small

Nightingale

areas of southern and eastern England, where it is often scrubby habitats that now remain their strongholds

It's a sad state of affairs given that humans have had such an infatuation with them over the centuries. From Chaucer to Shakespeare, the Nightingale's vocal skill has earned it a place in human history forever more. Throughout its literary mentions, it is often a female bird wrongfully cast as the owner of the voice that blesses European ears in the spring when the birds return from Africa. Her song was an expression of sorrow, and it was said that she would lean against a thorn whilst singing to infuse ever more grief into her notes.

SKULKING SONGSTER

Actually seeing a Nightingale is a difficult task, given their love of dense thickets, and even a loudly singing bird can be impossible to catch a glimpse of. If you're lucky enough to get a good view of a Nightingale, then you'll see why it was their vocal talents that made them famous and not their appearance. Slightly larger than a Robin, the Nightingale is a plain brown bird, other than its rufous tail.

BRING THE NOISE

The Nightingale's songs are incredibly complex, with over 200 different types of songs recorded, containing a selection of whistles and trills from across the melodic spectrum. Another defining feature of the Nightingale's song is the volume at which it sings. Males have been recorded singing at 95 decibels, about as loud as a chainsaw operating a metre away from your head.

One of the most iconic recordings of a Nightingale dates back to 19 May 1942. Broadcasting live from a Surrey garden, BBC Radio was planning to bring the sound of Nightingales singing in the still evening air into people's homes. However, this was 3 years into World War Two and, as the Nightingale sang, it was joined by the hum of 197 RAF bombers flying overhead on their way to Germany. It resulted in an extremely poignant recording, the sound of war juxtaposed with the beauty of nature's finest birdsong.

STARS OF THE SCREEN

A Nightingale nest has only featured on *Springwatch* once, given their rarity and their habit of nesting low down in dense undergrowth. At RSPB Minsmere in 2016, a nest low down in the nettles was watched as chicks were raised without any drama. On the final day of filming, there was suddenly a commotion at the nest as the adult birds appeared, clearly agitated, with wings raised and uttering scolding calls. The chicks hunkered into the nest but the cause of the alarm was unclear until an adder raised its head in the bottom of the screen, clearly making its way to their home.

As the venomous viper reached the nest, one of the adult birds dive-bombed the snake, delivering a peck to its head. The well-developed chicks bolted for cover, 'panic-fledging' as a means of escape. The adult Nightingales continued to stand their ground until the adder beat a retreat. It is these kinds of extraordinary interactions that the *Springwatch* cameras allow us to see, and it's believed to be the first time that anti-predator behaviour such as this has been filmed.

WHERE TO SEE: *Southern and eastern England. Hotspot counties include Suffolk, Sussex and Kent, with spots such as RSPB Minsmere, the Knepp Estate and Blean Woods offering you the chance to hear their legendary vocal talents.*

WHEN TO SEE: *As a migratory species, Nightingales are only in the UK from April through to early autumn. Your best chance to see them is when males first arrive back, when they're at their loudest and singing the longest before they pair up with a female!*

Thrushes

Thrushes might conjure up the classic image of a brown bird with a speckly chest. Whilst that is true for our Mistle Thrush and Song Thrush, their family also includes the humble Blackbird and migrants like Redwing, Fieldfare and Ring Ouzel. All are united by being on the larger side of the songbird spectrum and possessing long, sturdy legs evolved for spending lots of time on the ground foraging for food. As a family, they're arguably our finest songsters too and their rich, uplifting notes are ones we're all able to hear thanks to their ubiquity across Britain.

Blackbirds

A stalwart of almost any garden up and down our isles is the Blackbird. Watching them hop purposefully across our lawns, listening intently for the sound of worms beneath the soil, before furiously plunging their beaks into the turf to extract their juicy prize is a scene familiar to most.

The Blackbird is a bird of woodland originally but has found our human adjustments to habitat quite to its liking. Our gardens, with their open patches of lawn and borders with exposed soil, provide the perfect place to search for food whilst our hedgerows and shrubs are ideal spots for nesting.

The male is jet black in colour, with a striking yellow eye ring and beak packed full of carotenoids, which demonstrate his health and fitness to the females. The name 'Blackbird' is a misleading one, however, and is only representative of 50 per cent of the population, as females are brown all over with a varying amount of faint speckling on their breasts.

ON THE MENU

Blackbirds pop up repeatedly in folklore, particularly in song and rhyme. One of their most well-known appearances can be found in 'Sing a Song of Sixpence', where only three lines in we find 'four and twenty Blackbirds, baked in a pie.' Many interpretations of the rhyme have been proposed, including one that the Blackbirds are an allusion to monks during the period when Henry VIII was dissolving the monasteries.

One of the most curious lines in the rhyme is 'when the pie was opened, the birds began to sing.' It might seem impossible for any birds contained within a pie to still be remotely capable of singing a tune, but there was an unusual fad in the sixteenth century for making pies that could contain unharmed live birds. The idea was that they would fly out upon the pie being cut, to the shock and amusement of the guests around the dinner table.

CHRISTMAS CLASSIC

One of the most famous Christmas songs of all time also contains a hidden reference to the bird: 'The Twelve Days of Christmas'. You may be struggling to recall where exactly in the song the Blackbird appears – but our 'calling birds' that are given to us by our true love on the fourth day of Christmas are Blackbirds. In the original renditions of the song, this line was 'four colly birds' an old English nickname for Blackbirds and a reference to the male's black, coal-coloured plumage. The line has been changed over the years to become the 'calling' birds that we all sing today, but historically it was the humble Blackbird that had its place on this exclusive Christmas gift list.

LAST TO LEAVE

The Beatles sang of the 'Blackbird singing in the dead of night'. Whilst midnight may not quite be accurate, it is true that Blackbirds are one of the first and last birds to sing in the chorus. Their large eyes in relation to their head mean they can be active in very low light levels, and it's often their rich, lyrical phrases that are drifting from the aerials as the light creeps away.

WHERE TO SEE: *Blackbirds are common across Britain. The easiest place to see them is in parks and gardens.*

WHEN TO SEE: *Any time of year!*

Ring Ouzels

To find our only thrush that migrates to the UK specifically to breed, we have to travel to the upland landscapes of Britain. Here, the Ring Ouzel makes its home amongst the rocks.

Also known as the Mountain Blackbird, they arrive from North Africa's Atlas Mountains in the spring to breed. They're much larger than the Blackbird, although very similar in appearance but for the white 'necklace' that adorns their chest.

WHAT'S IN A NAME?

The word 'Ouzel' is a funny one. Historically, it was used to describe the Blackbird, hence 'Ring Ouzel' was used to differentiate this rarer species. Dippers were even thrown into the mix and given the name 'Water Ouzel', due to their perceived similarity in body shape and colour scheme. It's only the Ring Ouzel that has still clung onto its name nowadays, and its song echoing over the valleys of some of our most famous National Parks is a treat for the ears.

STARS OF THE SCREEN

Springwatch featured a nest of Ring Ouzels when we were based in the Cairngorms in 2019. A camera rigged on a nest in the mountains watched as the parent birds kept their growing chicks well fed with the plethora of worms that were available thanks to the wet weather. What we didn't expect was the male to return to the nest with a Common Lizard on one occasion! After a bit of struggling, one very ambitious chick managed to swallow it whole and secured itself a very hearty meal.

WHERE TO SEE: *Upland National Parks hold many of the best populations. Look for them in places like Dartmoor, the Peak District, the Yorkshire Dales, Eryri and the Cairngorms.*

WHEN TO SEE: *They arrive in mid-March and spend the spring and summer on their breeding territories. In autumn they turn up in the lowlands as they begin their migration, often stocking up on berries or feeding out in open fields.*

Song Thrushes

Throstle, Thrushel, Thrushfield, Mavis – the Song Thrush has been known by many names down the centuries. We have always been drawn to those species with beautiful songs and the Song Thrush's fervent, effusive, relentlessly upbeat tune has captured our imaginations.

The Song Thrush is the archetypal 'thrush', with its characteristic brown back and spotted chest. They tend to stay more hidden than Blackbirds and that, coupled with a population decline of 50 per cent since 1967, makes them a harder bird to see.

MUSICAL MAESTROS

Learning their song is the best way to find them – and what a song it is. It can be tricky, as the Song Thrush uses many different notes to make up its melody. The key thing to listen out for is the repetition that Song Thrushes love to work into their songs. Yes, they may use many different notes, but they will very often repeat a melody multiple times before moving on to the next one. Robert Browning wrote in 1845, 'That's the wise thrush, he sings each song twice over.'

No Song Thrush tune is exactly the same, but key into the repetition and the phrases they like to use, and you'll learn there are more around you than you thought. It helps that they blast out their song at max volume and spring evenings are often a good time to hear those notes come roaring out of the trees and filling the air with an irrepressible energy.

FAERIES' FRIEND

Song Thrushes are more than just a one-trick pony when it comes to their talents. These master musicians are excellent at crafts too, making a beautiful mud-lined nest in which they lay their piercing blue, flecked eggs. An Irish belief states that the faeries would make sure that the Song Thrush built its nest low down near their homes in the grass so that they would be able to enjoy the bird's song. If a thrush was found to have built its nest high up in a thorn bush, it was a sign that the faeries were unhappy and misfortune would come to the neighbourhood.

SMASH AND GRAB

They have also evolved a method for making use of a very specialised food source – snails. Snails are tricky food to eat given their shell, and getting them out of their fortified defences is impossible for most birds, but not the Song Thrush. Upon finding a snail, it will pick it up by the lip of the shell entrance with its beak and take it to a very particular spot. This is usually an exposed rock, but human objects like glass bottles and corrugated metal are also known to be used. These function as the Song Thrush's anvil, and it will waste no time bashing the shell against it with enough force to crack it open and retrieve the juicy mollusc meat inside. Granted, it may be a skill that's a little less sophisticated than their musical talents or clay work, but it gets the job done. And a good anvil can be hard to come by, so Song Thrushes have been shown to use favoured anvil spots for decades!

WHERE TO SEE: *Woodland or scrub is your best bet, but they can also be found in parks and gardens.*

WHEN TO SEE: *Learning their song in the spring is a great way to help find them and, like the Blackbird, they often start and finish the day with their powerful song.*

Song thrush

Mistle Thrushes

The Mistle Thrush is the heavyweight of the family, sometimes known as the Bull Thrush, and it isn't afraid to throw its weight around. They're named after their love of mistletoe berries, which at one time were believed to only germinate if they passed through the bird, but they're not particularly fussy and most fruit-bearing trees will satisfy their tastes throughout the autumn and winter.

FOOD FIGHT

Unlike the roaming flocks of Fieldfare and Redwing, Mistle Thrushes take a different strategy to ensure they're not going to run out of food throughout the leaner months. After finding a tree laden with berries, they will defend it vigorously – chasing off any intruders that dare to even so much as glance the way of their precious haul. Its tenacity at defending its patch earned it a Welsh name *Pen y Llwyn*, translating to 'master of the coppice', and a Mistle Thrush tree in winter is a guaranteed front-row seat for some drama.

STORM SINGER

The Mistle Thrush is undoubtedly a hardcore bird. They are one of the first songbirds to begin singing, as early as December in the depths of midwinter, as they begin to set up their territories. An old name for Mistle Thrush is Jeremy Joy, believed to have evolved from 'January Joy' in reference to their knack of raising the spirits and heralding the first signs of spring.

Their song is melancholic, almost wistful, and proclaimed from the tops of the highest trees they can find. Such is their commitment that, even in the driving rain and high winds that winter can throw at them, they remain stoically serenading. It's little wonder then that another name for the Mistle Thrush is Stormcock.

Thomas Hardy was so enamoured with them that he immortalised them in his poem 'The Darkling Thrush', published in 1900. Although the bird is described as 'small', the description of a thrush singing his 'full-hearted evensong' at Christmastime means it's generally identified as a Mistle Thrush.

WHERE TO SEE: *Found across most of Britain, look for them in woodland, parks and gardens. Keep an eye on berry bushes in the colder months to see if they're being guarded by a hungry Mistle Thrush.*

WHEN TO SEE: *Winter is the best time to see them in action. As well as defending food sources, they also begin to sing from the tops of trees as spring approaches, making them more noticeable.*

Redwings and Fieldfares

Every autumn, our numbers of thrushes are boosted by the annual Viking invasion of two species from the north. Redwings and Fieldfares breed in Scandinavia, but as the temperature drops and the days shorten, they take their leave and set out across the North Sea. Once in Britain, no berry or apple is safe as they raid hedgerows up and down the land and pillage orchards of their fallen bounty.

CARRIED ON THE WIND

The Redwing is the smallest thrush regularly found in Britain, coming in around the same size as a Starling. It has a beautifully marked face with a creamy 'supercilium', a fancy word for an eye stripe, that makes it easily recognisable if you get a good view. The red of a Redwing isn't actually on the wing itself. It's more of a reddy, orange smudge underneath the wing that creeps down the flanks, giving it the appearance of being caught trying to smuggle a ketchup bottle out of a supermarket. In parts of Scotland, they're known as Windthrushes, given their arrival on the winds of autumn.

Like many species of birds, Redwings do much of their migration at night, allowing them to avoid detection by hungry avian predators. Around a million of them are estimated to migrate to the UK each year. One of the great treats of autumn evenings is being able to stand outside on still nights and hear the soft 'tseep' call rain down from the dark sky as the Redwings pass over unseen above.

BIGGER BROTHER

The Fieldfares that arrive at the same time as the Redwings are the largest of the thrushes found in Britain. They're much bigger and are strikingly patterned with a mix of slate greys, deep browns, warm tans and those characteristic thrush speckles down their chest.

They are far more raucous than their politer, smaller cousins, announcing their presence with harsh 'tchak tchak tchak' calls as they fly over. 'Fieldfare' literally means 'the traveller over the field', given their roaming winter habits.

WHERE TO SEE: *They are found across Britain throughout autumn and winter. Hawthorn hedgerows laden with berries are a good place to spot them when they first arrive in the autumn. By mid-winter, orchards are a favourite as they feast on fallen apples. By the end of winter, when fruit is scarce, look for them feeding on open fields.*

WHEN TO SEE: *Look out for both species as they maraud across the British countryside from October to March.*

Warblers

If there is a group of birds that we must be most thankful towards for bringing us the delights of the dawn chorus as spring hurtles forward at a quickening pace, then it is the warblers. Their return to the UK from further south fills the warming air with a bountiful mix of joyful songs, a lyrical blur of scratchy, reeling, punchy, liquid notes from the varied songsters that perform them.

Chiffchaffs

The first to raise its voice as it celebrates the downfall of winter is the Chiffchaff. Its bouncy two-note song 'chiff-chaff-chiff-chaff-chiff' rains down from the bare canopy, spring's messenger sent to lift our spirits. It's a great species to start learning about bird song from as it begins its song before many other birds have fully started singing, meaning it's easy to isolate in a chorus. It's also a very simple song that says its own name, and you won't find many more obliging birds than that!

Blackcaps

The next warbler to add its song to the chorus is the Blackcap. This is a much more complex, piping song that bursts from the brambles. John Clare wrote poems celebrating the Blackcap, one entitled 'The March Nightingale', and they have been referred to as both Mock Nightingales and Northern Nightingales as a favourable comparison to the species most often seen as the champion of song. In autumn, our breeding Blackcaps depart for southern Europe and are replaced by Blackcaps from Central Europe that have cottoned on to the fact that our garden bird feeders and milder British climate have created a comfy place to spend the winter.

Sedge and Reed Warblers

As spring gathers pace, April begins to fill with the sounds of the warblers arriving back to the UK from their winters in sub-Saharan Africa.

In our wetlands, the Sedge and Reed Warblers find their voice, giving the normally eerily quiet reedbeds a constant hum of scratchy whistles and chatter. The Sedge Warbler sings faster, like a DJ constantly remixing at breakneck speed, whereas the Reed Warbler sings at a steady pace with more regular intervals. Both birds will sing at night, giving them old names like Night Warbler for the Reed and Night Singer or Night Sparrow for the Sedge.

The Reed Warbler makes an extraordinary nest, expertly woven low down amongst the reed stems. As the reed grows, the nest rises with it, carrying its precious cargo further clear of the water below. Building their nest like this keeps it safe from most predators, although their arch nemesis, the Cuckoo, seems to have no problem targeting Reed Warbler nests to lay its eggs in.

Reed warbler

Grasshopper Warblers

One of the most curious sounds the warblers bring to the spring chorus is that of the Grasshopper Warbler. This is an incredibly shy and skulking bird that, if it has intentions not to be seen, will make itself impossible to see. They creep through the grasses and brambles, scurrying on the ground like a little feathered mouse in the most unassuming way. It is often only when they sing that their presence becomes obvious, although to hear it without any prior warning, a bird is perhaps the last thing you would guess it to be!

As the name suggests, the Grasshopper Warbler makes a chirrupy song but not in short bursts like that of a cricket or grasshopper. Instead, it churns out the noise with a great persistence, like the reeling of a fishing rod going on indefinitely. It's such a strange sound in that it is both loud and quiet at the same time, often not changing in the volume the closer you get to the bird. Naturalist Gilbert White wrote of it in 1768:

> *Nothing can be more amusing than the whisper of this little bird, which seems to be close but though at a hundred yards distance; and, when close at your ear, is scarce any louder than when a great way off.*

Whitethroats

In hedgerows up and down the length of our isles, there is a bird-song that's one of the quintessential sounds of a summertime walk through the British countryside. It is a punchy, short, scratchy song, sometimes accompanied by a short display flight, that tells you a Whitethroat is near.

The Whitethroat's winter quarters in the Sahel sometimes suffer from long periods of drought, which affect the Whitethroat's over-winter survival. In the late 1960s, there was a huge crash in numbers when a severe dry spell saw a decline of 90 per cent in UK Whitethroat numbers. Even to this day, their population is still recovering.

Their widespread nature across the country has given them a range of folk names. Their love of the dense undergrowth led to Nettle Creeper, Nettle Bird and Hedge Chicken. It's their choice of grass as a nesting material that gave us Hay Jack, Strawsmall, Winnel Straw and Hay Tit, whilst their white throat feathers meant Beardie, Muffit and Whittie Beard were used.

Willow Warblers

Arguably one of the most beautiful songs of spring comes cascading down from the treetops from mid-April, like a babbling brook of liquid notes tumbling over the rocks. It is the sweet song of the Willow Warbler, one that has been heard right across the UK for decades. Francesca Greenoak called it 'the most abundant and widely distributed warbler in Britain' when writing in the late 1970s, but its fortunes are rapidly shifting.

Widespread losses have occurred of Willow Warbler numbers throughout southern Britain, with areas of perfect habitat emptying of their previously healthy populations. Evidence is building to suggest that it is climate change that has caused these parts of the UK to be too warm for the Willow Warbler's specific needs. As their range shifts north, millions of people have now lost the delights of its springtime song.

It is a warning that many of our birds, no matter how apparently common, are all at potential risk from climate change and its unseen consequences. Species that migrate long distances, like many of these warbler species, have that risk compounded by relying on conditions in different locations across the globe.

WHERE TO SEE: *Other than the shifting populations of the Willow Warbler, all of these species are widespread across the UK. Look for Reed and Sedge Warblers in reedbeds and the scruffy edges of wetland habitats, whilst Chiffchaffs and Blackcaps can be found in most woodlands and parks. Whitethroats love hedgerow edges and patches of scrub, as does the Grasshopper Warbler, which prefers a more open scrubland landscape.*

WHEN TO SEE: *Most of these birds arrive from April and it is their songs that make them such a delight. Look for them as they return to set up territories in early spring before they fall silent once they've raised their chicks. Chiffchaffs and Blackcaps now regularly overwinter in large numbers and can be seen at any time of year.*

Tits

The tit family undoubtedly claim a spot when it comes to the nation's best-loved birds.

There are officially six species of the tit family that we get in the UK, with some of them being more familiar than others. All of them will visit bird feeders, which means that, depending on where you are in the country, you'll be able to encounter at least one or two species pretty regularly. They're characterful birds, full of endless, seemingly inexhaustible energy and agile acrobatics. It's no wonder we love them so much.

Before we start, let's address the elephant in the room, shall we? Where does *that* name come from?

The origins of the word 'tit' go back a long way. It's the shortened form of the word 'titmouse', which is the name they were given in the fourteenth century. The 'mouse' part of the name is derived from a word that was given to small birds and animals, whilst 'tit' also meant something small. Over time, the 'mouse' bit was lost and they just became 'tits', paving the way for endless puns and double entendres!

Blue Tits

If *Springwatch* was to ever build a 'Mount Rushmore' of animals – those that have formed the backbone of the programme over its two-decade long run – then there's no doubt that Blue Tits would have a good claim to be first to have their face carved into the mountain.

Blue Tits are about as *Springwatch* as the theme tune. They've been almost ever present since the inception of the programme and we've followed the fates and fortunes of countless nests over the years. We root for the runts of the brood and hold our collective breath when the Great-spotted Woodpeckers come knocking for a tasty chick-sized snack.

Our love for Blue Tits comes from the familiarity we have with them. They're one of our brightest-looking birds, with a colour scheme that wouldn't look out of place in the canopy of the Amazon rainforest. Not only do they have a flamboyant dress sense, but they know how to flaunt it too. Their eye-catching acrobatic exploits on feeders make them a joy to watch and a bird that's pretty hard to miss.

ON THE UP

Blue Tits are bucking the trend for many bird species and have actually enjoyed a meteoric rise in numbers over the last few decades, with the latest population figures putting their number at well over 3 million territories. Part of this is down to their adaptability, meaning that they're able to live in most habitats as long as there's a smattering of trees. Whether that be woodlands, farmland, parks or gardens, as long as there are insects for them to feed on and holes for them to make their nests in, they'll be happy.

Blue tit

HUMAN HELPING HANDS

Another reason for their dramatic increase is down to our love for our garden birds. In the UK we provide around 150,000 tonnes of food for our feathered friends, which helps many of them through the cold winters when wild food becomes naturally scarce.

It's not only winter where we help them out either. As *Springwatch* shows, Blue Tits are very welcome receivers of nest boxes and will readily set up home in them. The last national census showed there were a minimum of 4.7 million nest boxes in private gardens across the UK – that's almost one for every pair of Blue and Great Tits in the whole country!

We've recorded thousands of hours of Blue Tit footage across the series and watched hundreds of chicks fledge the nest. Yet despite them featuring so regularly and competing with a cast of more unusual species, the Blue Tits remain a firm fan favourite all these years later!

However, despite their abundance and regularity on *Springwatch* since the beginning, more recently we've found them missing from our roster.

The long, cold spring of 2021 meant trees were late to come into leaf. The caterpillars that feed on the leaves were delayed too and so weren't around for when the Blue Tits needed to feed them to their chicks. This caused a lot of nests to fail and led to our first-ever series without a Blue Tit nest after almost 20 years. The year 2022 then followed suit but for entirely different reasons, with the incredibly warm spring meaning that they'd all fledged by the time the show was live!

WHERE TO SEE: *Parks, gardens and woodlands across the UK. There really aren't many places where you can't find a Blue Tit!*

WHEN TO SEE: *Any time of year is good to see these common birds. A nest box in a garden stands a good chance of attracting a pair in spring as long as there are some nearby trees.*

Great Tits

The bigger, bolder, badder member of the family. Just as widespread as their Blue cousins, they're easily distinguished by their striking black head and brilliantly white cheeks. They're also much larger – about the size of a Robin.

For this reason, they're one of the most dominant birds that come to feeders and can often be seen aggressively posturing, fanning their wings and opening their beaks in the direction of any lesser mortal they believe unworthy of sharing the spoils.

From our human perspective, it's difficult to appreciate the size differences at play here and how, for many of our other garden favourites, a Great Tit is really not to be messed with. Let's take your classic bird feeder brawl scenario of Blue Tit vs Great Tit. The size difference between the two species would be the equivalent of you turning up to a restaurant and having to fight off a brown bear to get to your table.

EVOLUTION IN ACTION

So devoted are Great Tits to bird feeders that it's become genetically encoded into their DNA. A study published in 2017 by researchers at Oxford University, Sheffield University, the University of East Anglia and the Netherlands Institute of Ecology showed that Great

Tits in the UK have a beak that's on average 0.3mm longer than their counterparts in the Netherlands. The scientists concluded that the likely reason for this was to help them reach into bird feeders to get at more food as bird feeding is a lot more prevalent in the UK than it is on the Continent.

SURPRISING TASTES

It's not only bird feeders that they eat from though. Great Tits are an incredibly adaptable species and almost nothing is off the menu. From their standard diet of seeds, berries and a huge range of invertebrates, Great Tits have also been recorded feeding on slightly more unusual things.

Extraordinary observations from Continental Europe have seen Great Tits entering caves to systematically search through cracks in the rocks to hunt pipistrelle bats as they awake from hibernation. There are also multiple accounts of them actively hunting other songbirds, with records of them doing their best Sparrowhawk impression and taking down birds such as Goldcrests and Redpolls.

BIRDSONG

It's well worth listening out for their song as one of the early heralds of spring. The males, with their glossier cap and thicker breast band, will take up a suitable singing post on exposed branches and belt out their simple two-syllable song. It's often described as though they're shouting 'teacher, teacher', although I've always found it sounds more like a squeaky wheelbarrow being pushed down a garden path!

STARS OF THE SCREEN

Springwatch 2018 featured Plucky, a runt-of-the-brood Great Tit chick who had a VERY narrow escape on its first venture into the big, wide world. Plucky left the nest with the rest of the siblings, despite the fact it wasn't able to fly yet. Instead of flying to the safety of cover, Plucky sat begging for food at the nest entrance, exposed on the small stump that the nest had been built in. All this noise got the attention of a hungry Jay, who came diving in to pluck Plucky from his perch. Thankfully, the little Great Tit lived up to its name and dived back into the nest chamber, squishing itself as far in as it could go, only millimetres away from the Jay's outstretched beak! Plucky left the nest a few hours later and hopped off into the undergrowth, leaving us all to imagine the story was to have a happy ending … although Chris Packham was less convinced!

WHERE TO SEE: *Wherever you find Blue Tits, their bigger cousin isn't likely to be far away!*

WHEN TO SEE: *Listen out for that early spring song from February or keep an eye on any garden bird feeders!*

Long-tailed Tits

First things first, Long-tailed Tits don't belong in this section. They technically deserve their own as the only member of bushtit family (*Aegithalidae*) to be found in the UK. In fact, they're actually more closely related to warblers like Chiffchaffs and Blackcaps than they are to Blue and Great Tits!

Regardless of which branch of the avian tree they're hanging upside down from, there's no doubt that they're one of the most adorable birds in existence. There's nothing these birds do that *doesn't* radiate an incredible amount of cute.

Let's start with the way they look. A tiny body with a tail sticking out that's just as long, combined with their short, stubby bill and big eyes, make them look as though they've flown right off the screens of an animated movie.

WHAT'S IN A NAME?

Long-tailed Tits have a wealth of different folk names linked to their appearance or behaviour:

Mumruffin

Bottle Tit

Bum Barrel

Bum Towel

Oven Bird

Hedge Jug

Jack-in-a-bottle

DID YOU KNOW?
Victorian ornithologist William MacGillivray was so keen to understand just how many feathers were used in a single Long-tailed Tit nest that, after the birds had finished with it, he dissected one and counted 2,379 feathers!

TRILLING TITS

Long-tailed Tits constantly communicate with each other as they move through the trees, excitedly trilling away. Listening out for them is a really good way to spot Long-tailed Tits and you can often

Long-tailed tit

hear a whole group of them moving up a hedgerow towards you before they burst onto the scene.

Another tick in our checklist of adorableness comes when you realise *why* they're so vocal – they're incredibly sociable birds. Winter is when they're at their most obvious, moving around in flocks that can sometimes reach up to 50 birds. Their constant calling is to keep the flock together as they roam around looking for food and keeping an eye out for predators. At night, these large flocks will huddle up together to keep each other warm and in spring, they'll even help each other out when it comes to feeding each other's chicks.

NIFTY NESTS

Long-tailed tits make one of the most exquisite nests in the avian world. Think of a Fabergé egg made out of moss, lichen and spider's webs and you're not a million miles from the truth. It's an extraordinary domed structure that's most often buried within the dense protective thorns of plants like hawthorn or bramble. Despite not having a single opposable thumb between them, a pair of Long-tailed Tits can weave together this amazing construction in a couple of weeks.

The walls of the nest are made predominantly with moss, held in place by spiders' webs, before the outside is decorated in flakes of lichen. This not only helps glue the whole structure together even further but also helps to camouflage it. Once the exterior work is done, it's onto the interior décor. Long-tailed Tits aren't ones to go for the minimal Scandi-chic aesthetic when it comes to furnishings, they go full throttle to pack their nest with as many feathers as possible.

They seem to love the large, plump feathers of birds like Woodpigeon and Pheasant and they rarely travel far from the nest as they search them out. It's impressive how many they actually manage to get their beaks on considering the limited time and space that they're operating in and it's here where they find themselves with an unlikely ally. Goshawks and Sparrowhawks, so often the nemesis of a little snack-sized Long-tailed Tit, suddenly become a worthwhile bird to have in the vicinity when piles of feathers are required!

Once the nest is built and lined, this extraordinary piece of construction has one more trick up its sleeve. Up to 15 eggs can be laid in a single nest, although 6 to 9 is more likely, and after a couple of weeks these begin to hatch. Over the next 2 weeks, these chicks will grow until they're ready to fledge and the issue now facing the nest is that it's about to get very cramped, very quickly. So, for its final party piece, the Long-tailed Tit deploys a trick that many human parents spend thousands of pounds on carrying out in the face of a growing family – the home expands. Thanks to its mossy walls stitched together with spiders' webs, the nest is able to stretch to accommodate the chicks growing inside until they're ready to leave the nest and enter the big wide world.

WHERE TO SEE: *Anywhere! They're widely spread across the country but particularly like scrubby areas with bramble and gorse to build their nests.*

WHEN TO SEE: *Most easily spotted in the winter months when large flocks roam the countryside – often descending onto fat balls and peanut feeders in gardens!*

Goldcrests and Firecrests

The title of the joint smallest bird in the UK has two holders: the Goldcrest and the Firecrest.

TINY TWOSOME

These two birds are impossibly small, with an average weight the same as a 20p coin at about 5g. Given their size, seeing either species can be difficult, but Goldcrests are actually a very common species throughout Britain. Firecrests play the role of their much rarer cousin and have only been breeding here since the 1960s, but have been expanding northwards ever since.

WHAT'S THE DIFFERENCE?

Given their similar size and behaviours, telling them apart can be difficult. Both of them have the eye-catching stripe on the top of their heads that is yellow in females and contains a flash of vibrant orange in males. Goldcrests are more plainly marked, however, especially around their face, which possesses a cute, doe-eyed appearance. Firecrests have opted for a fiercer, bolder look with striking facial markings giving them a white-and-black eye stripe, which proves the most reliable way to separate them.

IMPRESSIVE VOYAGE

Whilst the breeding populations of these species are relatively sedentary, many more arrive in the winter to escape colder temperatures in Continental Europe and Scandinavia. This is particularly true for Goldcrests, which can turn up in huge numbers (known as 'falls') on autumn days on the East Coast. It's incredible to think of a bird that weighs about the same as an A4 piece of paper having the ability to cross the tumultuous North Sea during the autumn.

HITCHING A RIDE

So scarcely believable was their feat of crossing the sea that it led to a particularly imaginative folklore story to explain how these birds managed to make it here against the odds.

At a similar time of year, Britain also sees the arrival of huge numbers of Woodcocks, who are fleeing the frozen grounds of northern Europe. Here is how we end up at an old name for the Goldcrest – the 'Woodcock Pilot'. Some sources state that to cross the sea, the Goldcrest would hop onto the back of the much larger Woodcock and steer it over the waves to its wintering grounds.

Perhaps a more realistic interpretation of events is found in the timings of both species' migrations. Given that hunters have always had a keen interest in Woodcocks as gamebirds, they would await the autumn arrival eagerly. The main influx of Goldcrests happens slightly before that of Woodcocks, and it may be that Goldcrests were seen as a sign that the Woodcocks were on their way, perhaps even testing – 'piloting' – the route to ensure that the bigger birds had safe passage!

STARS OF THE SCREEN

Chicks in a Goldcrest nest we featured in the Cairngorms in 2019 had an explosive introduction into the world. We'd been watching the nest for a while and the chicks were around fledging age but still hadn't made the jump. As the light faded from the day and the chicks seemed all snuggled down, they suddenly all burst from their nest in a blur of wings. It wasn't until the bush started shaking and the nose of a pine marten poked into the empty nest just seconds later that we realised what had caused them to leave! By scattering into the bushes, it's likely that some of them were able to survive and not end up on the Marten's menu that night.

WHERE TO SEE: *Goldcrests are found all over Britain, with Firecrests restricted to the southern half of England and Wales. Both species are particularly drawn to coniferous woodlands and it's always worth loitering around a yew tree to see if you can hear a high-pitched call ringing out of the dark foliage!*

WHEN TO SEE: *They're at their loudest during spring but in autumn they join mixed-species flocks, roaming through woodlands and scrub with other small birds where they can sometimes be seen more exposed.*

Goldcrest

Corvids

Mischievous, mystical and much maligned, we've had a complex and powerful relationship with the crow family for centuries. Their resourcefulness and intelligence, coupled with the diversity in the niches that they fill, mean that we see them in all habitats, such as the Carrion and Hooded Crows swaggering down the streets of our towns and the Ravens soaring over our highest peaks. From the call of Choughs echoing against sea cliffs to the screech of the Jay amongst our ancient oak woods and the Rooks in their tree-top towers, corvids are everywhere, for us all to enjoy in their brilliance and with their endless charisma.

Magpies

Perhaps no bird carries as much superstition with it into the modern day as the Magpie. You may well be one of the many people who still greet any Magpie they see with a tip of the hat, a salute or a polite greeting, asking how its day is going. A more extreme reaction to seeing a Magpie is recorded in Shropshire, where you were required to shout, 'Devil, Devil, I defy thee! Magpie, magpie, I go by thee!' and spit on the ground three times to ward off its wicked ways.

Encountering a Magpie was once seen as bad news, as they were almost universally regarded as an omen of evil. In Scotland, they were believed to carry a drop of the devil's blood beneath their tongue and one seen near the window of a house was a foreteller of death.

Magpies are still widely associated with being predictors of the future. Perhaps you still count the Magpies you see, always hoping for there to be more than the dreaded single bird:

One for sorrow,
Two for joy,
Three for a girl,
Four for a boy,
Five for silver,
Six for gold,
Seven for a secret never to be told.

BLACK AND WHITE

It is Magpies that give us the word 'pied' as a descriptor for other animals with black-and-white colouring. Since the thirteenth century, the birds were known as Pies and it was three centuries later that the word 'pied' was recorded. Get a good look at a Magpie and you'll see that they are dressed in far more than a simple monochrome colour scheme. Their wings and tail are glossed with beautiful greens, purples and blues that catch the light to rival the beautiful feathers of any other British bird.

Magpie

SHINY STEALERS

Magpies are famed for their love of shiny things, to such an extent that people who take things or collect pretty objects may be called a Magpie. A study in 2015 by Toni Shephard et al. aimed to test the validity of these charges levelled at the Magpie and found that there was no additional attraction to shiny objects than any other. The scientists concluded that humans simply notice more when Magpies occasionally pick up shiny objects, while it goes unnoticed when Magpies interact with less eye-catching items.

WHERE TO SEE: *Other than in parts of Scotland, Magpies are a bird that takes no real difficulty to find!*

WHEN TO SEE: *Magpies are resident all year round, so can be seen easily at any time of year.*

Jackdaws

Sitting on the chimney pots in many a quiet country village will often be a pair of Jackdaws. They're birds of the open country, feeding out in fields but still needing tree holes, buildings or cliff faces to make their untidy stick nests in. The smallest of the crow family, they're recognised by their grey heads and piercing white eye.

COSY CORVIDS

Jackdaws are the only corvid with a light iris and it's thought to have evolved so they can see each other when poking their heads into potential nest holes to see if it's a free spot. Jackdaws are fiercely territorial and will protect their chosen nest site right through the year. They're so often seen in pairs because they're one of the few species that really, truly mate for life. Pair bonds are cemented by bouts of *allopreening*, where the birds take it in turns to preen each other's head and neck. In winter, they come together in large flocks to communally roost, putting on aerial displays that can rival Starling murmurations for their mesmerising, shape-shifting, flocking behaviour. Some roosts can reach tens of thousands of Jackdaws in the winter, a cacophony of chattering calls, which gives them their collective name: 'a clattering'.

WHAT'S IN A NAME?

Where the name 'Jackdaw' comes from is still not fully known but there are some theories. The Jackdaw was originally known as *ċēo* or *ċeahhe* in Old English, with the C at the beginning being pronounced as 'ch'. It was derived from the Jackdaw's call and eventually evolved into the word 'Chough'. With Jackdaws being known as Choughs, a similar-sized crow found in coastal Cornwall was given the name Cornish Chough to distinguish it. This crow, with its lipstick-red bill and legs, eventually took the name completely, becoming the Chough we know today.

Jackdaws had many other names in the past, one of them being Daws, potentially named after another of their calls. The name Jack is used in animal names to denote a smaller version, e.g. Jack Snipe, and may have come to be used to signify it as the smallest of crows.

WHERE TO SEE: *Jackdaws are widespread across almost all of the UK but are most often found in villages and towns surrounded by countryside. Buckenham Marshes on the Norfolk Broads is home to the biggest roost site.*

WHEN TO SEE: *Jackdaws are active all year round but are at their most spectacular when they gather in their thousands at their winter roosts.*

Jay

Jays

You're more likely to hear a Jay than you are to see one. If you've ever walked in the woods and heard a formidable screech ringing through the trees, then it is a Jay that is the culprit. Its call is echoed in its old names Devil Scritch and Scold, originating in Somerset, and *schreachog choille* in Gaelic – 'the screamer of the woods'.

THE FEATHERED FORESTER

Despite being our most flamboyant crow, with its warm pink tones clashing with its electric blue wing patches, it is also our most secretive. Jays prefer the cover of trees and are particularly associated with oaks. Throughout autumn, Jays collect and bury acorns to see them through harsher winter conditions. They can store thousands of acorns in a season, each in a different place, and their extraordinary corvid brain means they're able to remember the locations of a large number. Those that are forgotten will germinate to produce new oak trees, and so the Jay is an important forester in its own right.

In fact, the Jay is widely considered to be responsible for the northward spread of oak forests across Europe at the end of the Ice Age!

STARS OF THE SCREEN

Given their shy, retiring nature and wily intelligence, finding Jays' nests can be tricky. As such, they've only appeared on the *Springwatch* live cameras once, from the Sherborne Estate in Gloucestershire in 2017. This was a particularly precarious nest that had begun to slide out of the tree, hanging at a worrying angle and causing the nation to watch with their hearts in their mouths as they wondered whether the chicks would fledge or fall. There were four well-developed chicks and by the second morning of the series, one had already fallen from the nest. Thankfully it had managed to find itself a perch above the ground and was still being fed by its parents. Its siblings were more successful and, with no great wish to hang around and suffer the same fate, all three successfully fledged later that day!

WHERE TO SEE: *Broadleaf woodland is the Jay's stronghold but getting a good view can be difficult. Many city parks are now home to good numbers and these birds are often quite relaxed around people.*

WHEN TO SEE: *Although they can be seen any time, autumn is when they're at their most obvious as they search for acorns to store ahead of the oncoming winter.*

Ravens

The Raven is a mighty bird. Differentiated from the crows by their large size, they stand similar in stature and wingspan to a Buzzard. A combination of their jet-black colour, evocative call and keen intelligence has captivated humans throughout the ages. They are found all across the northern regions of the globe and have become interwoven with the folklore of all cultures that have lived alongside them.

NORSE TALISMANS

The Old Norse name for the birds was *hrafn*, and the God Odin was known as the Hrafnagud, 'the Raven God'. He had a pair of Ravens – Huginn (Old Norse for 'thought') and Muninn ('mind') – that would fly across the world, gathering information for Odin before returning to perch upon his shoulder and whisper to him everything that they had seen.

A famous Norse banner carried the image of a Raven. Anglo-Saxon tradition stated that the flag was pure white in times of peace, but when the Danes marched to war, it would bear the image of a Raven. If the Raven's wings drooped, then defeat awaited them, but if the Raven stood proud with wings outstretched, then they were destined for victory.

In parts of the West Riding of Yorkshire, children were told of the 'black Raven' that would come to fetch them if they misbehaved, thought by some to be a call back to a time when marauding Norsemen were storming England's lands under the Raven banner.

ILL OMENS ...

The Raven's taste for carrion meant they were a common sight in the aftermath of battles. Their presence around death led to them being seen as an ill omen and they were said to have the power to predict oncoming events. In Denmark, the appearance of a Raven meant the death of a pastor, whilst in some parts of France, wicked priests were said to become Ravens when they died.

In Somerset, anyone who heard the 'kronk' cry of a Raven three times would turn away and cross their fingers to prevent bad luck. In Lincolnshire, it was unlucky to hear their call, particularly over the left shoulder.

... OR NOT

Not all depictions of Ravens are sinister. In the Jewish, Christian and Islamic traditions, the Raven was one of the first animals that Noah released from the Ark to find land after the Great Flood. The Welsh believed prosperity was assured if a Raven sat on the roof of your house and that they could restore the eyesight of the blind if you were kind to them, a belief that is likely linked to gruesome observations of them pulling out the eyes of corpses. Archaeologists in Hampshire found skeletons of Ravens deliberately buried as a seemingly symbolic gesture. These positive beliefs may be derived from the Celtic legend that King Arthur returned as a Raven after he was struck down in battle.

Raven

PROTECTORS OF THE REALM

Ravens are famous for their association with the Tower of London. It is said that at least six Ravens must remain at the Tower at all times or the Tower, the monarchy and the safety of the country are at risk. It is not known exactly when the legend began, although it is said that King Charles II was the first to be warned of the dangers of removing the Ravens during his reign in the 1600s. A group of Ravens is unflatteringly called an unkindness, and nowadays a group of seven make up the Royal flock. They are no longer wild, free-flying birds and are kept under the watchful eye of the Ravenmaster, a Beefeater dedicated to their care.

REWARDING ROOSTS

Ravens are not often seen in large flocks, unlike other members of their corvid family. Adults will form strong pair bonds and maintain their territory, never straying too far, but young birds can be highly social. At Newborough Forest on the Isle of Anglesey, a winter roost of hundreds of adolescent, non-breeding Ravens comes together.

An ingenious study by Bangor University found that the birds meeting at the roost were using it as a chance to share information on where they were finding food. Researchers used sheep carcasses baited with colour-coded, digestible beads so that they could analyse the colour of regurgitated pellets found at the roost site. The study found that pellets of the same colour would begin to spread through the roost, emanating from the first bird to have found and fed on the food source. This was evidence that the Ravens shared the location of the carcasses with the birds around them.

BIRD BRAINS

Ravens are renowned for being one of the smartest animals on the planet. They have a brain that is amongst the largest of any bird and can solve complex problems when presented with them. When measuring their ability across a range of tasks to assess their intelligence, a recent study by Simone Pika et al. showed that they performed just as well as chimpanzees and orangutans. What's more, they reach this bird-brained level of genius at only 4 months old!

Ravens have been seen engaging in what appears to be play behaviour too. From sliding down snowbanks to playing with sticks, they've been observed doing many things that seem to serve no clear survival purpose. Ravens are also excellent mimics and are able to copy sounds from their environment. In captivity, they're capable of mimicry of a whole suite of different sounds, with many viral videos attesting to the fact that they're also adept at copying human speech.

WHERE TO SEE: *A recent surge in population has meant that Ravens are now much easier to see. Although there are some urban breeding pairs in cities like Bristol, our National Parks, where they can remain undisturbed, are some of the best places to see them.*

WHEN TO SEE: *Ravens are resident all year round but their joyous barrel rolling courtship display in late winter is a heartening early sign of spring.*

Starlings

There is no better British avian spectacle than winter's dance of the Starlings.

MURMURATION MAGIC

Get yourself to a murmuration site, whether it be a pier, a woodland or a reedbed, as the sun begins to set and wait. Soon, the cast will begin to arrive, slowly at first, in small pockets here and there. The small flocks circle, not wanting to commit to landing in the roost in such small numbers. Further small groups appear on the horizon, silhouetted against the sky. As they wheel round, they begin to coalesce, as if drawn together by magnets. As the light continues to drain, larger squadrons begin to arrive. They join the swirling throng, kicking up the volume of squeaks and clicks as the birds orchestrate their dance.

If you're lucky, you may soon be overawed by so many Starlings that you can't keep them all in one field of view. They continue to pour in from the surrounding countryside where they've been feeding all day, heading to these communal roost spots to spend the evening. The giant flocks begin to twist, pulsate and writhe like a giant super organism in the evening air – each individual Starling a cell in its movement.

SAFETY IN NUMBERS

Gatherings like this rarely escape the notice of predators for too long, and it's the appearance of a bird of prey, like a Peregrine, diving at the mega flock that adds the real crescendo to the piece. Like a shoal of fish, the Starlings flash one way then the other, contorting themselves into all manner of shapes never seen before and unlikely to be seen ever again. The sound of a million wings beating the air as the flock rolls and dives rushes through your ears like the wind.

TOUCHDOWN

The final phase comes when the birds decide it's time to go to roost. That's when this great black swirling tornado reaches down its spike to the ground and the Starlings begin to drain into the safety of their chosen location. It is as though someone has suddenly unplugged a bath, and within seconds the mass of Starlings in the sky can have disappeared, leaving nothing but the last glimpses of light and the sound of their excited chatter.

DID YOU KNOW?
Murmurations can occur almost anywhere, in towns, cities or out in the countryside. In 1949, so many Starlings once settled on the hands of Big Ben's clock tower that it ground the great clock to a halt!

BETTER OFF TOGETHER

Murmurations are astonishing things. The word itself relates only to flocks of Starlings and has its origins in the 1300s, probably derived from the sounds of the birds chattering. This hypnotising display is not only about safety in numbers from both predators and the cold, but the mass gatherings also allow them to exchange information gathered over the course of the day about which locations might present the best feeding areas.

STARS OF THE SCREEN

In *Winterwatch* 2021, we rigged cameras on the roost at Aberystwyth Pier to get a Starling's eye perspective of being inside the roost. As well as capturing a unique view of the display, we were also surprised to see a visiting Barn Owl, making nightly commutes in the pitch-dark to help itself to an easy meal.

COAT OF MANY COLOURS

Let us not overlook the Starling's singular charm though. Take a look at any individual Starling in spring, marching purposefully around a lawn looking for leatherjackets and other soil grubs. They possess an iridescence that drips off their feathers in shades of blue, green and purple. A vibrant yellow beak, with a blue base in males and pink in females, stands out in sharp contrast. In winter, they lose some of their sheen but retain their white spots that sit upon them like a constellation of stars.

MUSICAL MIMICRY

It's not just their sharp dress that impresses, as Starling vocalisations can be dizzyingly varied: a mix of electric chattering, buzzing clicks and whistles, with layers of mimicry of not only other bird songs but other sounds they've heard in the landscape. The more complex and impressive the song, the more it says about the singer, and so a male's striving to put himself above the rest means no two Starling songs are ever the same.

There are even reports of Starlings being able to pass sounds down through the generations. On the Isle of Coll in Scotland, Starlings living in the walls of a tumbled-down bothy were heard to mimic the sound of a two-stroke engine. Hidden in the undergrowth were the decaying remnants of just such an engine, out of use long before any of the currently occupying Starlings had been alive. It seems to be that, as the Starlings copy each other's songs throughout the years, the engine became a key feature to this very specific group of birds.

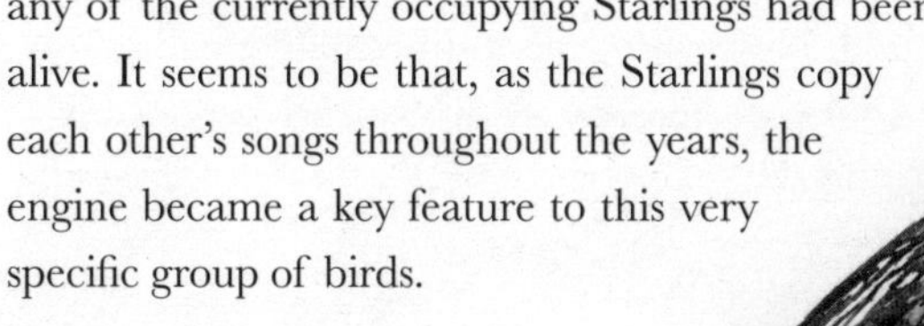

Starling

FAMOUS INSPIRATION

Mozart was so enamoured with the skills of one particular Starling that he bought it upon hearing it repeat a piece of one of his piano concertos in 1784. He kept it for 3 years, becoming hugely affectionate towards the bird. After its death, he organised a considerable funeral ceremony, including a procession and a substantial gravestone. It's claimed by some experts that elements of Mozart's 'A Musical Joke' bear similarities to the vocalisations of Starlings.

WHERE TO SEE: *Despite declining in recent decades, Starlings are still a widespread and common bird over much of Britain and they aren't hard to find. In some towns, often by the seaside, birds can be so tame they can be hand-fed.*

WHEN TO SEE: *Winter murmurations are one of the most spectacular things you can see in Britain. Regular ones can be found on the Somerset Levels and the piers of Aberystwyth and Brighton, but they can appear suddenly anywhere.*

Sparrows

There's perhaps not a bird that owes more to the actions of humans than the House Sparrow. Believed to have originally evolved somewhere in the Middle East, the House Sparrow followed in the footsteps of pioneering farmers as agriculture took over Europe.

The farmed landscape, with its mix of hedges, buildings and plenty of spilt grain, had all the ingredients for sparrows to thrive. When they reached the English Channel, the trade boats carrying goods back and forth between the British Isles and the Continent likely provided Sparrows the perfect means to cross – with some even suggesting it was at the time of the Romans that House Sparrows first appeared in the fossil record.

A BIRD OF MANY NAMES

Their long association with us, and their charismatic, bustling characters, have endeared them down the generations. They have many nicknames, some still commonly in use today. In Northern England they're often called Spugs or Spuggy, in Scotland Spur, Speug, Sprig or Spriggies. In the south of England, we get Sparr and the famous Cockney Sparrar. There are other names too – Spadger, Philip and Spatzie, all of which reflect the familiarity we have with them.

ON THE MENU

So abundant and readily available were sparrows that they were even eaten in the past! Special pots were used around areas where sparrows were nesting in order to catch them, and there's evidence of these pots, continued use until at least the 1800s. The caught sparrows would be baked into a sparrow pie or pudding, making an easily accessible source of meat for most people.

But it wasn't their apparent culinary uses that caused the sparrow to experience a frightening fall in numbers. Over the last 53 years, it's estimated that Britain has lost 22 million of its House Sparrows. Just why numbers crashed in urban areas is still a mystery, although there are some positive signs in recent years that the tide may be turning and sparrow populations are showing signs of recovery.

COUNTRY COUSINS

The fortunes are looking similar for the House Sparrow's much rarer cousin too. The Tree Sparrow, a smaller, daintier bird with a chestnut cap and black bib that's shared between the sexes, was always more of a rural, farmland species. Like many farmland birds, it suffered a huge decline in line with the intensification of agricultural practices – for every Tree Sparrow today, there were around 20 in the 1970s.

WORLD DOMINATION

That is not to say that sparrows are faring badly on a global scale. The House Sparrow, in particular, has followed us around the world. On top of their natural spread following the plough across Europe and Asia, they have also been introduced to North and South America, Southern Africa and parts of Australasia. It's estimated that the House Sparrow is the second most numerous bird on Earth with a population of around 1.4 billion.

It's something to think about next time you're watching them from the window hanging from your bird feeder or listening to them chattering loudly from a bush. They're a feathered reflection of ourselves, spreading out alongside us as we took them on our journey of modern civilisation. A plucky survivor that saw an opportunity and took it – and we need to protect its continued status as our right-hand bird.

WHERE TO SEE: *House Sparrows are best seen around, well, houses. Urban and suburban areas, villages and farmyards are all good places to find them. Tree Sparrows are much harder to spot these days. They're generally found on farmland in eastern and lowland areas of Britain, with East Yorkshire being a particular stronghold.*

WHEN TO SEE: *Any time of year.*

Finches

The finch family is a colourful group of small songbirds that can be found in all corners of the UK. In general, they're specialised at eating seeds of all different shapes and sizes, and so possess a variety of different-shaped beaks to match. Some of them are huge, like the powerful Hawfinch, capable of crushing cherry stones and exerting a force of 4kg per square centimetre with its powerful beak. There's a finch for most British habitats and many are regulars to our garden bird feeders. Chaffinches, Goldfinches and Greenfinches all come to bird tables, whilst Siskins and Redpolls will often frequent them too if you live in the right areas. In winter, Bramblings arrive from Scandinavia to add a splash of orange to our garden's cast of feathered friends.

Goldfinch

Goldfinches

Twittering, tinkling flocks of Goldfinches, their golden wings catching the light as they turn into the sun, are a much more common sight nowadays than they were in the past. Between 1995 and 2020, numbers rose by 156 per cent as they took advantage of the bird food we provide in our gardens. They're particularly fond of Niger seed and sunflower hearts, both of which have become much more readily available in recent years.

CHARMING BEHAVIOUR

A flock of Goldfinches is called a 'charm', a rather lovely word that has its roots in the Old English *c'irm,* meaning the 'blended sound of many voices', which is reflective of their constant chirpy twittering as they move around together. Despite their delicate appearance, a Goldfinch feeding frenzy on the bird feeders can often be a riotous experience. There's lots of posturing and shouting as they muscle each other out of the way to secure access to food.

WONDROUS WEEDS

Goldfinches have evolved a slender beak made for teasing out small seeds from thistles or teasels and when they're not bickering over gourmet garden food, they're often roaming forgotten bits of land or field edges on the lookout for plants that many would class as 'weeds'. The Anglo-Saxons called them *thisteltuige*, meaning 'thistle-tweaker', whilst names such as Thistle Warp and Thistle Finch have survived through to the present day.

WHERE TO SEE: *Goldfinches are regulars to bird feeders containing Niger seed and sunflower hearts. Why not try attracting them to your own garden?*

WHEN TO SEE: *Goldfinches are a common bird all year round but look for them perched on teasel heads in the winter, extracting the delicate seeds.*

Greenfinches

The Greenfinch is a stocky member of the finch family that is widespread across much of the UK. Males have a distinctive green body with canary yellow wings and tail, whilst females are slightly duller, but all possess the chunky bill used for cracking larger seeds.

KILLING WITH KINDNESS?

Greenfinches have regularly visited garden bird feeders for decades, feeding on peanuts and a range of seeds. However, since 1967 their population has crashed by almost 70 per cent thanks to a disease that is passed between bird feeders and is particularly lethal to the Greenfinch. *Trichomonas gallinae* is a protozoan parasite that affects the back of birds' throats, causing them difficulty in swallowing and breathing. It first came to prominence in 2005 and, although other species of birds are affected, it is the Greenfinches that have been the hardest hit and are now red listed as a result of their severe decline.

The parasite is passed through contaminated food or water supplies and so it is important to regularly clean any feeders or bird baths to protect birds from disease transmission.

WHERE TO SEE: *Greenfinches are still widespread, although in much lower numbers. Parks, gardens and farmland remain the best habitats to find them in.*

WHEN TO SEE: *Spring is the best time to find them, when males can be tracked down by following their loud 'wheezing' song, which they belt out from the tops of trees. Familiarise yourself with the sound and it will make it much easier to spot them!*

Bullfinches

There is perhaps not a better male/female bird combination in the British Isles than a pair of Bullfinches. Both sexes have the smart black-capped head leading into a chunky, glossy black beak. Both have a glossy blue-black tail and flight feathers, with a pale bar that sits just above them and a white rump that's easily visible as they fly away. The male has a rich reddy-pink breast whilst the female is a warm buff colour. They look great together and, unlike most other songbirds, they form strong pair bonds that last throughout the year. They will often be seen in pairs no matter the time of year.

ORCHARD ENEMY

The name Bullfinch comes from their large head and stocky shape, also giving us Bull Spink, Bully and Bulldog in the past. Other names point towards its love of eating the buds of trees – Bud Bird, Bud Finch and Bud Picker – and it is this habit that has brought the species into conflict with humans. Bullfinches have a particular taste for the buds of fruit trees, which has made them a pest in orchards. An Act of Parliament in the sixteenth century put a bounty on their heads, offering one penny for 'everie Bullfynche or other Byrde that devoureth the blowth of fruit'.

WHERE TO SEE: *Bullfinches will come to gardens if there is enough cover to hide in but are generally found in areas with large hedgerows or well-developed scrubby areas. Listen out for their soft plaintive call as a way of detecting them hidden in the foliage.*

WHEN TO SEE: *Any time of year! Bullfinches are resident all year round, although may be more likely to visit gardens in winter.*

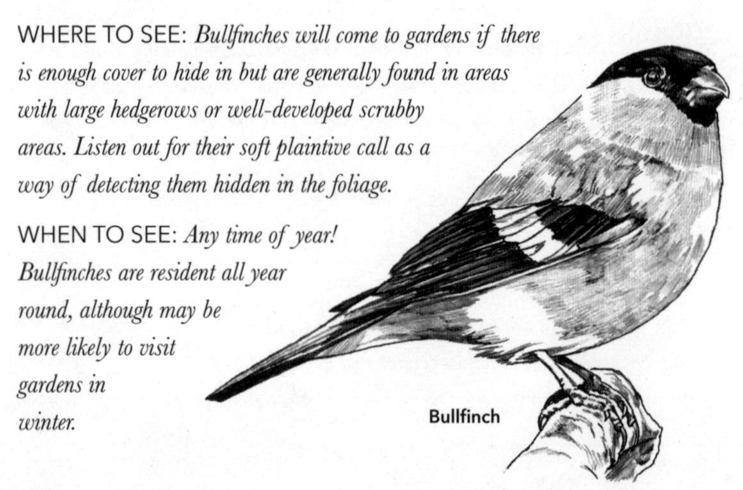

Bullfinch

Linnets

The Linnet is a bird of open country, farmlands and heathlands, where scattered cover provides them places to nest and seed plants provide them with plenty of food to eat. They love to nest in the spiny embrace of gorse bushes, especially early in the year before other bushes have come into leaf, and so have been known as the Gorse Thatcher, Thorn Linnet and Furze Bird.

CELEBRITY STATUS

The Linnet was a common bird of the countryside for hundreds of years and, like many finch species, was taken into captivity thanks to its appealing song. Its familiarity to us is reflected in the amount of times it is mentioned in popular culture throughout the ages, appearing in the works of Alfred Tennyson, Charles Dickens, Robert Burns, William Blake, William Wordsworth, the music of *Sweeney Todd* and even the Netherlands' entry into the 2014 Eurovision Song Contest!

FADING FAST

Sadly, the Linnet belongs to the suite of farmland birds that have suffered so much decline in recent decades, with numbers falling sharply since the 1960s. Its close relative, the Twite, is a bird of higher ground and coastal areas and has suffered even worse. It is facing imminent extinction as a breeding species in England and, although there is a small Scottish population, experts are concerned that they may soon share a similar fate.

WHERE TO SEE: *Farmland or heathland are still the best places to find Linnets. Check the tops of gorse and hedgerows for the best chances of seeing a male singing with his pink chest proudly puffed.*

WHEN TO SEE: *In winter they form big flocks, particularly on farmland that grows plants specially to produce winter seed for birds.*

Crossbills

The Crossbill has a name that does exactly what it says on the tin. The upper and lower mandibles of the beak are crossed at the tip, a specialised adaptation to allow them to feed on the cones of conifers. They work by gaining leverage to separate the scales of the cone, allowing them to use their tongue to get at the seed.

BLESSED BIRD

Religious legend tells that the Crossbill bent its beak trying in vain to extract the nails from the cross that Jesus Christ was crucified on. It is also here that the male bird got its red plumage, from the blood of Christ as the bird tried to assist him.

FOLLOW THE FOOD

Crossbills' lives are dictated by the cone crop and they will nest at any time of year if they find an area with a suitable bounty to support them. This can mean that, even in midwinter with the snow falling on her back, a female Crossbill can be found high up in a pine dutifully incubating her precious eggs. They are highly nomadic and huge numbers will sometimes descend upon the UK from the Continent if there has been a bumper breeding year or a failure of the cone crop that requires them to seek more food. These are known as 'irruptions' and have been written about for centuries, given that this unusual bird can appear in places it has rarely been seen before. Matthew Paris wrote the first-known record in 1251:

> *At the turn of the year, at the season of the fruits, certain wonderful birds never before seen in England appeared, particularly in our orchards. They were little bigger than larks and ate the pips of apples and nothing else from the apples. So they robbed the trees of their fruit very grievously. Moreover, they had the parts of the beak crossed and with them split the apples as if with pincers or a pocket knife.*

A BIRD TO CALL OUR OWN

Thanks to the forestry industry there is, these days, an abundance of conifer plantations spread around the UK, meaning that Crossbills are now permanent residents right down into southern England. In parts of northeast Scotland, there also lives the Scottish Crossbill. This is the only bird found in the UK and nowhere else in the world, although it is so similar to Common Crossbills (and the rarer Parrot Crossbill) that some debate whether it's truly a species in its own right.

WHERE TO SEE: *Given their nomadic nature, Crossbills really can turn up anywhere as long as the habitat is right. Large conifer plantations throughout the country are likely to host Crossbills, so keep your eyes trained at the tops of trees to catch a glimpse of them feeding silently away on the cones. Crossbills are particularly drawn to water, so puddles on forest tracks will often tempt them out of the canopy.*

WHEN TO SEE: *In winter you're more likely to see them flocked together, and this is often when irruptions are happening and numbers in the UK swell.*

Crossbill

Index